T0181953

Aircraft Flight Instruments and Guidance Systems

Written for those pursuing a career in aircraft engineering or a related aerospace engineering discipline, *Aircraft Flight Instruments and Guidance Systems* covers state-of-the-art avionic equipment, sensors, processors and displays for commercial air transport and general aviation aircraft.

As part of a Routledge series of textbooks for aircraft engineering students and those taking EASA Part-66 exams, it is suitable for both independent and tutor-assisted study and includes self-test questions, exercises and multiple-choice questions to enhance learning.

The content of this book is mapped across from the flight instruments and automatic flight (ATA chapters 31 and 22) content of EASA Part-66 modules 11, 12 and 13 (fixed/rotary-wing aerodynamics and systems) and Edexcel BTEC nationals (avionic systems, aircraft instruments and indicating systems).

David Wyatt has over 40 years' experience in the aerospace industry and is currently Head of Airworthiness at Gama Engineering. His experience in the industry includes avionic development engineering, product support engineering and FE lecturing. David also has experience in writing for BTEC National specifications and is the co-author of *Aircraft Communications and Navigation Systems*, *Aircraft Electrical and Electronic Systems* and *Aircraft Digital Electronic and Computer Systems*.

Aircraft Flight Instruments and Guidance Systems

Principles, operations and maintenance

David Wyatt

Routledge
Taylor & Francis Group

LONDON AND NEW YORK

First edition published 2015
by Routledge
2 Park Square, Milton Park, Abingdon, Oxon, OX14 4RN

and by Routledge
711 Third Avenue, New York, NY 10017

Routledge is an imprint of the Taylor & Francis Group, an informa business

© 2015 David Wyatt

British Library Cataloguing in Publication Data
A catalogue record for this book is available from the British Library

Library of Congress Cataloging-in-Publication Data
Wyatt, David, 1954–
    Aircraft flight instruments and guidance systems : principles,
    operations, and maintenance / David Wyatt.
        pages cm
    1. Aeronautical instruments. 2. Guidance systems (Flight) I. Title.
    TL589.W93 2014
    629.135—dc23   2014001225

ISBN13: 978-0-415-70683-4 (pbk)
ISBN13: 978-1-315-85897-5 (ebk)

Edna Margaret Wyatt
23 September 1930 – 17 December 2012

Contents

Preface

This book forms part of the aircraft maintenance books series; it can also be read as a standalone item, either in parts on its entirety. For continuity purposes there are cross-referenced overlaps with other titles in the book series; where the cross reference is relatively small in detail, the text and/or diagrams are repeated in this book. Where the cross reference is relatively large in detail, the text and/or diagrams are not repeated in this book.

This book contains new and updated reference material based on the latest sensors, processors and displays. It provides a blend of theory and practical information for aircraft engineering students. The book includes references to state-of-the-art avionic equipment, sensors, processors and displays for commercial air transport and general aviation aircraft.

The content of this book is mapped across from the flight instruments and automatic flight (ATA chapters 31, 22) content of EASA Part 66 modules 11, 12 and 13 (fixed/rotary wing aerodynamics, and systems) and Edexcel BTEC nationals (avionic systems, aircraft instruments and indicating systems).

To be consistent with EASA definitions, this book adopts the following terminology:

- 'Aeroplane' means an engine-driven fixed-wing aircraft heavier than air that is supported in flight by the dynamic reaction of the air against its wings.
- 'Rotorcraft' means a heavier-than-air aircraft that depends principally for its support in flight on the lift generated by one or more rotors.
- 'Helicopter' means a rotorcraft that, for its horizontal motion, depends principally on its engine-driven rotors.
- 'Aircraft' means a machine that can derive support in the atmosphere from the reactions of the air other than the reactions of the air against the earth's surface. In this book, 'aircraft' applies to both aeroplanes and rotorcraft.

Many of the subjects covered in the aeroplane aerodynamics and automatic flight control chapters can be applied to rotorcraft, e.g. aerodynamic drag and interlocks respectively. Any common subjects are not duplicated.

Any maintenance statements made in this book are for training/educational purposes only. Always refer to the approved aircraft data and applicable safety instructions.

Acknowledgements

- Aspen Avionics
- Aviation Today
- Avidyne Corporation
- CMC Electronics Inc.
- Cool City Avionics
- Dassault Aviation Group
- Embraer
- Gama Aviation
- Garmin
- Mid Continent Instruments
- Northrup Grumman

- Royal Aeronautical Society
- Journal of Aeronautical History Volume 1, Paper No. 2011/1: 'RAE Contribution to All-Weather Landing', Sir John Charnley CB, MEng., FREng., FRIN, FRAeS

Cover image: Dassault Aviation Falcon 7X, featuring fly-by-wire (FBW) technology, a subject covered in more detail in this book.

Images and reference material remain the property of each respective organization.

1   The earth's atmosphere

This chapter describes the environment in which aircraft fly. The reader will gain an understanding of three important features of the earth's atmosphere: air pressure, temperature and density. The underpinning theory for these three characteristics is explained to provide the basis of aircraft flight instruments and guidance. The chapter concludes with a review of the international standard atmosphere.

ATMOSPHERIC COMPOSITION

The atmosphere surrounding the earth extends up to approximately 250 miles, 400 km or some 1,320,000 feet. For practical aircraft engineering purposes, we are only concerned with the lower parts of the atmosphere. The approximate chemical composition of the earth's atmosphere can be expressed as a percentage by volume:

* Nitrogen   78%
* Oxygen   21%
* Others*   1%

Water vapour is found in varying quantities up to a height of 26,000 to 30,000 feet (8 to 9 km). The amount of water vapour in a given mass of air depends on the temperature of the air and whether or not the air has recently passed over large areas of water. The higher the temperature of the air, the higher the

amount of water vapour it can hold. Thus at altitude where the air temperature is least, the air will be dry, with very little water content.

The earth's atmosphere, as seen in Figure 1.1, can be divided into five principal zones. These zones are considered as concentric layers. Starting with the layer nearest the surface of the earth, these are known as the:

* Troposphere
* Stratosphere
* Mesosphere
* Thermosphere
* Exosphere.

Of the total atmospheric mass, approximately 75 per cent is concentrated within the troposphere; approximately 25 per cent is in the stratosphere. There is no

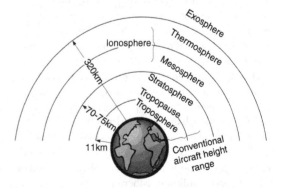

*Others include carbon dioxide, hydrogen, neon and helium

1.1 Principal layers of the earth's atmosphere

'top' to the atmosphere; it just becomes very thin at greater heights. Nonetheless, even at 200 miles above the earth's surface, there are still sufficient molecules of air to make spacecraft or meteors glow with heat due to friction. The boundary between the troposphere and stratosphere is known as the tropopause and this boundary varies in height above the earth's surface from about 25,000 feet (7.5km) at the poles to 59,000 feet (18km) at the equator. An average value for the tropopause in the 'International Standard Atmosphere' (ISA) is around 36,000 feet (11km). The thermosphere and the upper parts of the mesosphere are referred to as the ionosphere, since in this region ultraviolet radiation is absorbed in a process known as photo-ionization. In the above zones, changes in temperature, pressure and density take place. From an aerodynamic perspective, only the troposphere and stratosphere are significant.

TEST YOUR UNDERSTANDING 1.1

What are the three layers of atmosphere nearest the surface of the earth?

The tropopause is between the troposphere and which other layer?

The ionosphere includes which regions of the upper atmosphere?

GASES

In the study of gases we need to consider the interactions between temperature, pressure and density (remembering that density is mass per unit volume). A change in one of these characteristics always produces a corresponding change in at least one of the other two.

Unlike liquids and solids, gases have the characteristics of being easily compressible and of expanding or contracting readily in response to changes in temperature. Although the characteristics themselves vary in degree for different gases, certain basic laws can be applied to what is called a perfect gas. A perfect, or ideal gas is one which has been shown (through experiment) to follow or adhere very closely to these gas laws. In these experiments one factor, e.g. volume,

is kept constant while the relationship between the other two is investigated. In this way it can be shown that the pressure of a fixed mass of gas is directly proportional to its absolute temperature, providing the volume of the gas is kept constant.

ATMOSPHERIC PRESSURE

The air surrounding the earth has mass and is acted upon by the earth's gravity, thus it exerts a force over the earth's surface. This force per unit area is known as atmospheric pressure. At the earth's surface at sea level, this pressure is found by measurement to be $101,320 \text{ N/m}^2$ or in Imperial units 14.7 lbf/in^2.

Outer space is a vacuum and is completely devoid of matter, consequently there is no pressure in a vacuum. Therefore, pressure measurement relative to a vacuum is absolute. For most practical purposes it is only necessary to know how pressure varies from the earth's atmospheric pressure. A pressure gauge is designed to read zero when subject to atmospheric pressure, therefore if a gauge is connected to a pressure vessel it will only read gauge pressure. So to convert gauge pressure to absolute pressure, atmospheric pressure must be added to it.

KEY POINT

Absolute pressure = Gauge pressure + Atmospheric pressure

PRESSURE MEASUREMENT

Devices used to measure pressure will depend on the magnitude (size) of the pressure, the accuracy of the desired readings and whether the pressure is static or dynamic. Here we are concerned with barometers to measure atmospheric pressure and the manometer to measure low pressure changes, such as might be encountered in a laboratory or from variations in flow through a wind tunnel. Further examples of dynamic pressure measurement due to fluid flow will be encountered later in this book with aircraft pitot-static instruments.

The two most common types of barometer used to measure atmospheric pressure are the mercury and aneroid types. The simplest type of mercury barometer is illustrated in Figure 1.2. It consists of a mercury-filled tube which is inverted and immersed in a reservoir of mercury. The atmospheric pressure acting on the mercury reservoir is balanced by the pressure (ρgh) created by the mercury column. Thus the atmospheric pressure can be calculated from the height of the column of mercury it can support. The schematic mechanism of an aneroid barometer is shown in Figure 1.3. It consists of an evacuated aneroid capsule, which is prevented from collapsing by a strong spring. Variations in pressure are felt on the capsule that causes it to act on the spring. These spring movements are transmitted through gearing and amplified, causing a pointer to move over a calibrated scale.

A common laboratory device used for measuring low pressures is the simple U-tube manometer, as seen in Figure 1.4. A fluid is placed in the tube up to a certain level, and when both ends of the tube are open to atmosphere the level of the fluid in the two arms is equal. If one of the arms is connected to the source of pressure to be measured it causes the fluid in the manometer to vary in height. This height variation is proportional to the pressure being measured.

The magnitude of the pressure being measured is the product of the difference in height between the two arms Δh, the density of the liquid in the manometer and the acceleration due to gravity, i.e. pressure being measured *gauge pressure* $= \rho g \Delta h$.

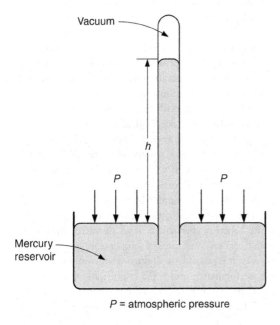

1.2 Simple mercury barometer

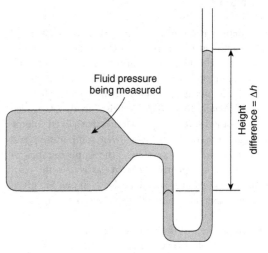

1.4 Simple U-tube manometer

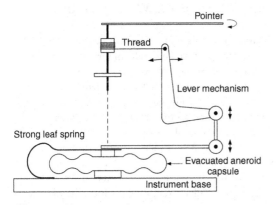

1.3 Aneroid barometer schematic

KEY POINT

Air density is proportional to atmospheric pressure.

Air density is inversely proportional to absolute pressure.

TEMPERATURE

Temperature is a measure of the quantity of energy possessed by a body or substance. It measures the vibration of the molecules that form the body or substance. The sun heats the earth and its surrounding atmosphere through radiation. These molecular vibrations slow down as temperature decreases, and only stop when the temperature reaches absolute zero, at minus 273.15 degrees Celsius. As the atmosphere warms up, the air expands and rises through convection. As the air rises, it cools and there is a gradual decrease in temperature through the troposphere.

The method used to measure temperature depends on the degree of hotness of the body, or substance. Measurement devices include liquid-in-glass thermometers, resistance thermometers and thermocouples.

DENSITY

Air density is the mass per unit volume of the air in the earth's atmosphere. Air density decreases with increasing altitude, as does air pressure. It also changes with variation in temperature or humidity.

Humid air is less dense than dry air; this was established by the Italian physicist Amadeo Avogadro in the early 1800s. Avogadro found that a fixed volume of gas in a container, at a given temperature and pressure, would always have the same number of molecules irrespective of what gas was in the container. As previously stated, perfectly dry air in a given volume contains mainly nitrogen and oxygen.

- 78 per cent of the air is nitrogen molecules, each with a molecular weight of 28 (2 atoms, each with an atomic weight of 14).
- 21 per cent of the air is oxygen, with each molecule having a molecular weight of 32 (2 atoms, each with an atomic weight of 16).

Avogadro discovered that, if water vapour molecules are introduced into dry air, some of the nitrogen and oxygen molecules are displaced (the total number of molecules in the given volume remains the same). The water molecules, which have now replaced the nitrogen and/or oxygen, each have a molecular weight of 18. (One oxygen atom with an atomic weight of 16, and two hydrogen atoms, each with an atomic weight of one.) The given volume of moist air is now lighter with added water vapour, compared with dry nitrogen/ oxygen. In other words, replacing nitrogen and oxygen with water vapour decreases the weight of the air in the given volume; its density has decreased.

INTERNATIONAL STANDARD ATMOSPHERE

Due to different climatic conditions that exist around the earth, the values of temperature, pressure, density, viscosity and sonic velocity (speed of sound) are not constant for a given height. The International Standard Atmosphere (ISA), as seen in Figure 1.5, has therefore been established to provide a standard for calculating and monitoring aircraft performance, and for the calibration of aircraft instruments.

The ISA is a hypothetical atmosphere based on average values, meteorological and physical theory. The performance of aircraft, their engines and rotor blades, or propellers, is dependent on the variables quoted in the ISA. It will be apparent that the performance figures quoted by manufacturers in various parts of the world cannot be taken at face value but must be converted to standard values using the ISA. If the actual performance of an aircraft is measured under certain conditions of temperature, pressure and density, it is possible to deduce what would have been the performance under the conditions of the ISA. It can then be compared with the performance of other aircraft, which have similarly been adjusted to standard conditions.

PROPERTIES OF AIR WITH ALTITUDE

Temperature falls uniformly with height until about 36,000 feet (11km). This uniform variation in temperature takes place in the troposphere, until a temperature of 216.7 K is reached at the tropopause. This temperature then remains constant in the stratosphere, after which the temperature starts to rise once again.

The ISA value of pressure at sea level is given as 1013.2 mb. As height increases, pressure decreases, such that at about 16,400 feet (5km) the pressure has fallen to half its sea-level value and at 49,200 feet (15km) it has fallen to approximately a tenth of its sea-level value.

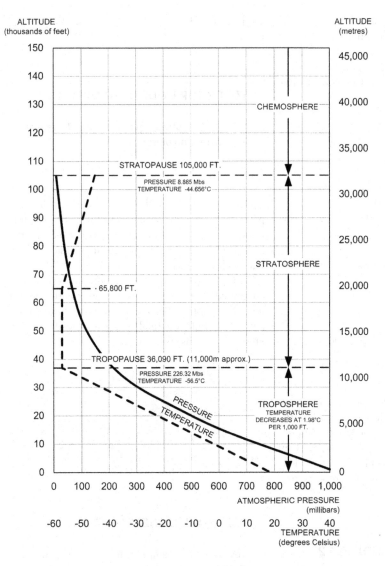

ALTITUDE
(thousands of feet)

ALTITUDE
(metres)

1.5 The International Standard Atmosphere (ISA)

The ISA value of density at sea level is 1.225 kg/m^3. As height increases, density decreases, but not as fast as pressure – such that, at about 32,700 feet (6.6km) the density has fallen to around half its sea-level value and at about 59,000 feet (18km) it has fallen to approximately a tenth of its sea-level value.

Humidity levels of around 70 per cent water vapour at sea level drop significantly with altitude. Remember that the amount of water vapour a gas can absorb decreases with decrease in temperature. At an altitude of around 59,000 feet (18km) the water vapour in the air is approximately 4 per cent.

KEY POINT

Humid air is lighter, or less dense, than dry air.

The conditions adopted have been based on those observed in a temperate climate at a latitude of 40 degrees North up to an altitude of 105,000 ft (32km). Based on certain simple assumptions, the primary variables in the atmosphere (pressure, temperature and density) are in fairly good agreement with observed annual values. The standard atmosphere is therefore an arbitrary set of conditions that are accepted as the basis of comparison for aviation engineering and science. The sea-level values of some of the more important properties of air, contained in the ISA, are:

- Temperature 288.15 °K or 15.15 °C
- Pressure 1013.25 mb or 101,325 N/m^2
- Density 1.2256 kg/m^2
- Speed of sound 340.3 m/s
- Gravitational acceleration 9.80665 m/s^2
- Dynamic viscosity 1.789 × 10–5 N s/m^2
- Temperature lapse rate 6.5 °K/km or 6.5 °C/km
- Tropopause 11,000 m, 56.5 °C or 216.5 °K

Note the following Imperial equivalents, which are often quoted:

- Pressure 14.69 lb/in^2
- Speed of sound 1120 ft/s
- Temperature lapse rate 1.98 °C per 1000 ft
- Tropopause 36,090 ft
- Stratopause 105,000 ft

Note that the temperature in the upper stratosphere starts to rise again after 65,000 feet at a rate of 0.303 degrees Celsius per 1,000 feet or 0.994 degrees Celsius per 1000 metres. At a height of 105,000 feet (32km) the chemosphere is deemed to begin. The chemosphere is the collective name for mesosphere, thermosphere and exosphere.

TEST YOUR UNDERSTANDING 1.2

Describe what happens with temperature in the upper stratosphere.

The speed at which sound waves travel in a medium is dependent on the temperature and the bulk modulus (K) of the medium concerned, i.e. the temperature and density of the material concerned. The denser the material, the faster is the speed of the sound waves. For an aircraft in flight, the Mach number (M), named after the Austrian physicist Ernst Mach, may be defined as the aircraft speed divided by

the local speed of sound in the surrounding atmosphere.

TEST YOUR UNDERSTANDING 1.3

Describe how atmospheric pressures decreases with increasing altitude.

The reader should have already met static and dynamic pressure in the study of fluids. (If not, then this is addressed in *Aircraft Engineering Principles* (AEP). Look back at the energy and pressure versions of the Bernoulli's equation.)

Fluid in steady motion has both static pressure energy and dynamic pressure energy (kinetic energy) due to the motion. Bernoulli's equation showed that, for an ideal fluid, the total energy in a steady streamline flow remains constant.

KEY POINT

Static pressure energy + Dynamic (kinetic) energy = Constant total energy

---

TEST YOUR UNDERSTANDING 1.4

Describe how temperature decreases with altitude.

In particular, with respect to aerodynamics, the dynamic pressure is dependent on the density of the air (treated as an ideal fluid) and the velocity of the air. Thus, with increase in altitude there is a decrease in density; the dynamic pressure acting on the aircraft as a result of the airflow will also drop with increase in altitude. The static pressure of the air also drops with increase in altitude.

Compared to temperature and pressure, humidity has a small effect on air density.

MULTIPLE-CHOICE QUESTIONS

1. Temperature decreases uniformly with altitude until about:

   (a) 36,000 feet (11km)
   (b) 65,000 feet (20km)
   (c) 105,000 feet (32km)

2. The ease with which a fluid flows is an indication of its:

   (a) Humidity
   (b) Temperature
   (c) Viscosity

3. Temperature in the upper stratosphere starts to:

   (a) Decrease after 65,000 feet
   (b) Increase after 65,000 feet
   (c) Increase after 36,000 feet

4. For an ideal fluid, the total energy in a steady streamline flow:

   (a) Remains constant
   (b) Increases
   (c) Decreases

5. An average value for the tropopause in the International Standard Atmosphere is around:

   (a) 105,000 feet (32km)
   (b) 65,800 feet (20km)
   (c) 36,000 feet (11km)

6. The amount of water vapour that a gas can absorb:

   (a) Decreases with increase in temperature
   (b) Decreases with decrease in temperature
   (c) Increases with decrease in temperature

7. The three layers of atmosphere nearest the surface of the earth are known as the:

   (a) Stratosphere, chemosphere, troposphere
   (b) Chemosphere, troposphere, stratosphere
   (c) Troposphere, stratosphere, chemosphere

8. With increasing altitude, the atmospheric pressure decrease will be:

   (a) Non-linear
   (b) Linear
   (c) Highest in the stratosphere

9. The boundary between the troposphere and stratosphere is known as the:

   (a) Tropopause
   (b) Stratopause
   (c) Ionosphere

10. Compared with dry air, humid air is:

    (a) Heavier, or less dense
    (b) Lighter, or less dense
    (c) Lighter, or more dense

2 Air-data instruments

Air-data (or manometric) instruments provide the pilot with key information needed for the basic operation of an aircraft's altitude and airspeed. These parameters are derived from measurements of the earth's atmosphere. This chapter describes the various instruments used to achieve these basic measurements, together with additional data that can be derived. Before delving into the various air-data instruments and systems, some generic principles of instrumentation are given.

The chapter sets out to encompass the various air-data instruments and systems installed on a range of aircraft, from smaller general aviation through to larger public-transport aircraft. Technology used across this range of air-data instruments and systems includes simple electromechanical devices through to solid-state sensors and electronic displays.

INSTRUMENTATION

The principles of instrumentation apply to all aircraft instruments, not just air-data instruments and systems. Subsequent chapters in this book will describe flight-control systems that use air-data sensors. The subjects of navigation and engine instrumentation are covered in two other titles in this book series, *Aircraft Navigation and Communication Systems* (ACNS) and *Aircraft Electrical and Electronic Systems* (AEES). Although the term 'instrument' used in this book is within the context of an electromechanical device with an analogue display, the terminology can be applied to solid-state sensors and displays.

KEY POINT

The terms used in association with instrumentation are: accuracy, precision, sensitivity, calibration, error, correction, tolerance, hysteresis and lag.

Accuracy is used to determine truth and comparison with a true value, e.g. temperature. An instrument is deemed to be accurate if it displays a value that agrees with the actual conditions, e.g. a thermometer should read 100 degrees Celsius in boiling water at sea level. The inaccuracy of an instrument is subjective and depends on the context. Comparison with a given parameter requires periodic checks to establish accuracy compared with a known value under test conditions.

KEY POINT

The aircraft engineer needs to recognize the inaccuracy of an instrument, either via pilot reports and/or through maintenance manual limits.

Precision is used to determine exactness and repeatability. A precision instrument will give exact indications on a repeatable basis. To illustrate this feature, take two similar thermometers used to measure the temperature of boiling water. When taking the first measurement, both thermometers give an indication of 102 degrees Celsius, i.e. an error of 2 degrees (or 2 per cent); both thermometers have the same accuracy. On a subsequent measurement, the first thermometer reads 101 degrees Celsius and the second reads 102 degrees Celsius. The first thermometer has higher accuracy, the second has higher precision.

> **KEY POINT**
>
> A precision instrument has repeatable results with constant test conditions.

Sensitivity is the ratio of output to input, and is a measure of the smallest signal the instrument can detect. To illustrate this, consider two sets of scales used to measure weight, kitchen scales and a vehicle weighbridge. The first is used to weigh relatively small amounts, maybe up to 1kg; the second is used to measure relatively large amounts, maybe up to 50,000 kg. Whilst the kitchen scale could measure an amount of 50g when placed on the weighbridge, the indication would be small if not zero. For the same measured amount of 50g, the kitchen scales are more sensitive because the indication would be 50 per cent of the measurable range.

Calibration of an instrument is the method of indicating the units of measure on the display. For example, a pressure gauge might display pounds per square inch (lbs/in^2) or kilograms per square centimetre (kg/cm^2). To ensure that instruments are retaining their reliability, they need to be range tested against a known standard on a periodic basis. To calibrate an instrument, it is adjusted to obtain the required accuracy.

Instrument error is the difference between the true reading and the observed reading. Errors are stated as positive when the reading is higher than the true reading and negative when the reading is lower than the true reading. Errors are generally quoted as a percentage of the true reading at the given observed reading.

Correction of a measurement is the amount needed to adjust the error to obtain the true reading. For examples of temperature and weight corrections, see Table 2.1.

The tolerance of an instrument is the acceptable error that is allowed for continued performance of its function. An aircraft instrument should be designed, manufactured and maintained to an accuracy that is fit for purpose. Referring to Figure 2.1, tolerance of indications will normally be specified in one of three ways:

1. As a percentage of scale range
2. As a percentage of the observed reading
3. As a given amount either side of a scale reading; this may vary at various points on the scale.

> **KEY POINT**
>
> Tolerance is the permissible or allowable error of an instrument.

Hysteresis is normally associated with electromechanical instruments that are repeatedly exercised over the full range. As a result of wear on mechanical components, or internal friction, there can be a delay between the action and reaction of a measuring instrument. Hysteresis is the amount of error that results from this delay. Hysteresis can also be an inherent feature of the installation, e.g. the room temperature thermostat in a central-heating system. The thermostat utilizes a predetermined amount of

Table 2.1 Examples of temperature and weight corrections

Measurement	True reading	Observed reading	Error	Correction
Temperature	50 degrees	55 degrees	+5 degrees	−5 degrees
Weight	50 grams	45 grams	−5 grams	+5 grams

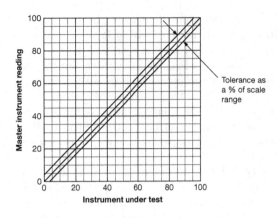

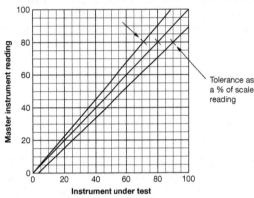

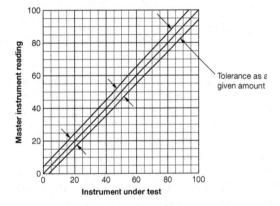

2.1 Tolerance indications

TEST YOUR UNDERSTANDING 2.1

Explain the difference between accuracy and precision.

TEST YOUR UNDERSTANDING 2.2

Give examples of instrument sensitivity.

TEST YOUR UNDERSTANDING 2.3

Explain the terms calibration, error, correction, tolerance.

ALTIMETERS

Overview

The aircraft's altimeter is essentially a calibrated barometer; it detects changes in atmospheric pressure as the aircraft climbs or descends and gives a visual indication of barometric height. The location of an altimeter in a traditional instrument panel is shown in Figure 2.2. The altimeter (top-right) is one of six primary flying instruments. The arrangement of these instruments is described in more detail in a subsequent chapter. A basic mechanical altimeter display and schematic is illustrated in Figure 2.3. The various components are housed in a sealed case which is connected to the external still air (static) atmospheric pressure via an internal static pipe and external static vent, or port. The altimeter consists of an evacuated

hysteresis (or lag) to maintain the desired temperature. Too little hysteresis would result in constant switching between on/off heating cycles; too much hysteresis would mean large temperature differences between cycles.

Attitude indicator
(artificial horizon)

Air speed
indicator

Altimeter

Turn
coordinator

Vertical
speed
indicator

Heating indicator

2.2 Primary flight instruments

triple-stack diaphragm that is mechanically coupled to pointers that are displayed over a suitably calibrated dial. Three aneroid capsules are used to give the altimeter more sensitivity. The mechanism is compensated for temperature variations by a U-shaped bimetallic bracket attached to the diaphragm. The barometric scale is visible through an aperture in the dial; the scale is adjusted via a knurled knob. Adjustments to the barometric scale are needed to compensate for changes in local atmospheric pressure (more on this later).

As the aircraft climbs, and atmospheric pressure falls, the aneroid diaphragm or capsules in the altimeter expand. A very small linear motion is transmitted to the gear mechanism, via the rocking shaft and a connecting link, which move the gears and pointers to show the altitude. As the aircraft descends, and atmospheric pressure increases, the aneroid capsules expand.

ICAO Q codes

The International Civil Aviation Organization (ICAO) Q codes are a standardized collection of three-letter message codes; they were initially developed for radiotelegraph communication, and later adopted by other services including aviation. Altimeters must be set and corrected to known atmospheric pressures; this is of great importance to the safe operation of the aircraft in all phases of flight. Referring to Figures 2.4 and 2.5, three settings are used in accordance with ICAO Q codes and associate terminology:

• QNH: altimeter referenced to mean sea-level atmospheric pressure so that the altimeter reads altitude based on local atmospheric pressure (may be either a local, measured pressure or a regional forecast pressure (RPS). When QNH is set on the altimeter, the displayed units (normally feet) are referred to as 'altitude'.
• QFE: altimeter referenced to local airfield pressure such that the altimeter reads zero on landing and take-off. When QFE is set on the altimeter, the displayed units (normally feet) are referred to as 'height'.
• QNE: altimeter referenced to ISA's standard atmospheric pressure at sea level; equal to 1013.25 mbar or hPa. When QNE is set on the altimeter, the displayed units are referred to as 'pressure altitude' expressed in flight levels. Flight levels are expressed in hundreds of feet, so 20,000 feet is flight level 200; 27,500 feet is flight level 275 and so on.

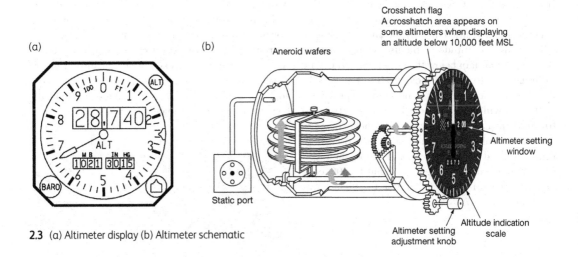

2.3 (a) Altimeter display (b) Altimeter schematic

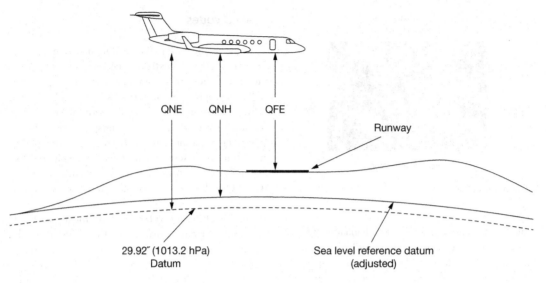

2.4 Q codes

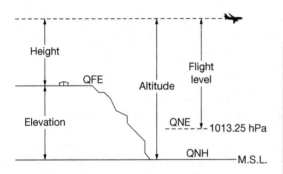

2.5 Altimeter terminology

Transition altitude (TA), as seen in Figure 2.6, is the altitude above mean sea level at which (under the direction of air traffic control) the aircraft changes from the use of altitude to the use of flight levels. When operating at or below the TA, aircraft altimeters are usually set to QNH. Above the TA, the aircraft altimeter pressure setting is normally adjusted to QNE. This will be the standard pressure setting of 1013 hectopascals (millibars) or 29.92 inches of mercury. In the United States and Canada, the transition altitude is fixed at 18,000 feet. In Europe, the transition altitude varies between 3,000 and 18,000 feet; air traffic control (ATC) will advise the flight crew accordingly.

Altimeter maintenance

> **KEY POINT**
>
> As with any maintenance statements made in this book, always refer to the approved aircraft data and applicable safety instructions; the following is for training/educational purposes only.

Altimeters normally do not require a great deal of maintenance once installed, other than visual inspection and a zero-reading check. When installing a new or repaired/overhauled altimeter, the instrument will have been fully checked by the manufacturer or repair shop. Typical on-aircraft checks include a zero-adjustment test, leak test and range test; details will be given in the aircraft maintenance manual.

> **KEY POINT**
>
> When carrying out a pressure-leak test on an altimeter, this will check the instrument case, not the capsule or accuracy. (Can you explain why?)

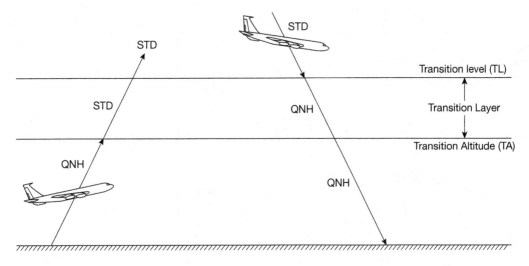

2.6 Transition altitude

VERTICAL-SPEED INDICATORS

Overview

(a)

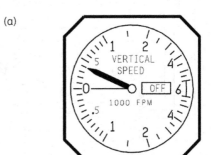

The vertical-speed indicator (VSI), or rate-of-climb (RoC) indicator, is based on the rate of change of atmospheric pressure; it is used to indicate the vertical component of the aircraft's speed. The VSI is normally located as shown in Figure 2.2; it indicates the rate of climb or descent by a single pointer that moves over a graduated dial. A VSI display and schematic is given in Figure 2.7; the VSI is normally graduated in thousands of feet per minute.

The mechanism is housed in a case that is air-tight; the case incorporates a pressure connection that is connected to still air (static atmospheric pressure) via a connector. The mechanism consists of a pressure-sensitive capsule, metering unit or choke (a 'graduated' or 'calibrated leak') and a gear mechanism to amplify the capsule's movement and drive the pointer. Still-air (static) pressure is applied to the inside of the capsule; under steady-state conditions this pressure will be the same as the internal pressure in the instrument case.

When the aircraft climbs, the atmospheric pressure reduces, resulting in a pressure difference between the inside of the capsule and the inside of the case. Static pressure is also supplied into the case, but this is through the metering unit so there is a delay in equalizing the pressure differential. Whilst the

(b)

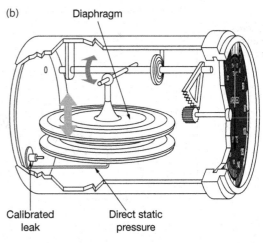

2.7 (a) VSI display (b) VSI schematic

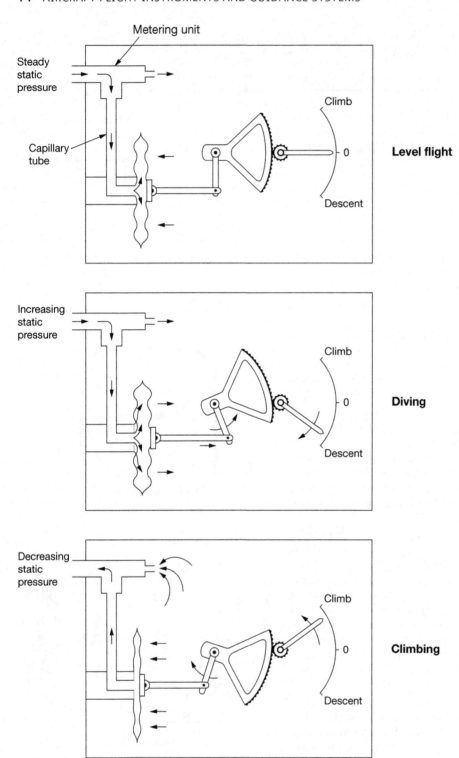

2.8 VSI scenarios

pressure differential exists, the pointer will indicate the pressure difference as the rate of climb. The amount of lag between the internal capsule pressure and the inside of the case depends on the aircraft rate of climb. When the pressure differential is reduced to zero, the pointer will indicate zero rate of climb. Figure 2.8 illustrates typical VSI scenarios during climb and descent.

TEST YOUR UNDERSTANDING 2.4

Explain what happens with a VSI when the aircraft descends.

Some VSIs are designed to sense vertical acceleration, as well as changes in air pressure. The instantaneous VSI (Figure 2.9) gives a more accurate display by eliminating lag time of the manometric VSI. The accelerometer is in the form of a piston that acts as a pump. When the aircraft begins a descent, the piston rises immediately in the cylinder, causing temporary increase of pressure to the capsule. When the initial acceleration decreases, i.e. during steady descent, the piston returns to its original position – by this time the differential pressure will have been established.

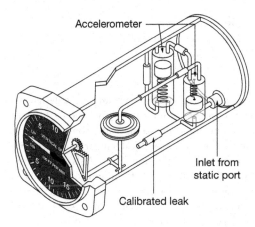

2.9 Instantaneous VSI

VSI maintenance

VSIs normally do not require a great deal of maintenance when installed, other than visual inspection and a zero-reading check. When installing a new or repaired/overhauled instrument, it will have been fully checked by the manufacturer or repair shop. Typical on-aircraft checks include checking that the indicator reads zero when the aircraft is on the ground and range test; details will be given in the aircraft maintenance manual.

KEY POINT

When testing a VSI through simulated rates of climb/descent, do not drive the indicator beyond its limits – this could damage the instrument.

AIR-SPEED INDICATORS

Overview

The air-speed indicator (ASI) displays the speed at which the aircraft is travelling through the air; it does not indicate speed over the ground. The ASI relies upon measuring the difference between the pressure created by movement through the air (dynamic or pitot pressure) and that of still air (static pressure). Pitot and static pressures are sensed via a pitot tube and static vent (or port). In some installations these are combined into a single pitot head. The typical ASI location is shown in Figure 2.2; a typical ASI display and schematic is illustrated in Figure 2.10.

Dynamic pressure

Air molecules in motion possess energy and therefore exert pressure on any object in their path. This dynamic pressure varies with air density and speed. If a volume of moving air is captured via an open-ended tube and stopped completely in a sealed capsule, the total energy of the air theoretically remains constant. The dynamic (or kinetic) energy is converted into

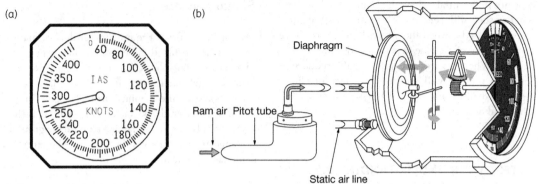

(a)

(b)

2.10 (a) ASI display (b) ASI schematic

pressure energy; for all practical purposes this can expressed mathematically as:

$$\text{Pressure} = \tfrac{1}{2}\rho V^2$$

Where ρ is the local air density, and V is the speed.

Principles of an ASI

The ASI is calibrated according to the law: dynamic pressure $= \tfrac{1}{2}\rho V^2$ where ρ is the standard sea-level air density and V is the indicated airspeed. With a pressure capsule connected to pitot (dynamic) pressure and static (still-air) pressure, movement of the capsule due to the sensed pressures is connected via a linkage mechanism to move a single pointer over a graduated scale.

With movement through the air, the pitot tube senses the total pressure comprising dynamic and static pressure; this total pressure is applied to the inside of the pressure capsule via a pipe connection. Static pressure is applied to the outside of the capsule.

KEY POINT

The ASI capsule is subjected to two opposing pressures, dynamic and static.

The total pressure sensed inside the capsule has the static pressure subtracted from it by the pressure sensed outside the capsule. The net result is that

movement of the capsule only occurs as a result of dynamic pressure. The instrument therefore indicates dynamic pressure in terms of speed. (ASIs typically indicate speed in knots or miles per hour.) The indicated air speed (IAS) varies as a function of the square root of the dynamic pressure.

ASI errors

Air-speed indicators are subject to errors that occur for various factors; these errors can be minimized to a certain extent to ensure accuracy and precision of the instrument. The errors arising from measurement of air speed are:

1) Instrument error
2) Position (pressure) error
3) Compressibility error
4) Density error

Instrument error occurs as a result of the design and construction of the unit; these give rise to different readings during calibration. Aside from mechanical wear, friction and so on, local temperature variations affect the ASI's calibration. Some ASIs incorporate bi-metallic compensators to minimize the effect of temperature on metal components, in particular the pressure capsule. As with all instruments, the ASI will have a specified tolerance allowed during manufacture, repair and on the aircraft.

Position (or pressure) error is caused by the sensed static pressure not being the true ambient still-air pressure. This is caused primarily by the location of the static vents, or ports. When an aircraft is first designed, a series of test flights is required to determine the optimum location for the pitot and static vent. These tests will be carried out over various speed, altitude and aircraft configurations. Actual results are compared with predicted results and (assuming that the ASI type and optimum pitot-head/static-port locations are acceptable) correction factors applied. These corrections can be in the form of tables and/or graphs for use by the flight crew and performance engineers. Aircraft with computerized equipment and systems could have these corrections built in via software.

> **KEY POINT**
>
> Position error on an ASI can be partially caused by air flow around the pitot head.

Compressibility errors occur because air is a gas, not a fluid; at higher air speeds, the air in the pitot tube compresses, causing the ASI to over-read. Air with low density is easier to compress than high-density air. At sea level, the dynamic pressure resulting from an air speed of 250 knots will give rise to a compressibility error in the order of 1–2 knots; at 35,000 feet, the same indicated air speed causes a compressibility error in the order of 20–25 knots.

> **KEY POINT**
>
> Higher pressure sensed as a result of compressibility causes higher ASI readings.

Density errors occur because ASIs are calibrated for sea-level air density. When flying at sea level, the indicated air speed is the true speed through the air. When flying at altitude, the indicated air speed is less than the true speed through the air. Although true air speed (TAS) is of more use for navigation purposes (more about this later), indicated air speed (IAS) is important for all flight phases since it represents the air loads on the airframe and is directly related to stall speed. An aircraft flying at 30,000 feet will have a true air speed of approximately 437 knots, and indicated airspeed of approximately 280 knots. Although the TAS is high, the air loads and stall speed will be the same as 280 knots at sea level.

Aircraft speeds

The speed of an aircraft can be expressed in several ways. Indicated air speed (IAS) is the reading of the air-speed indicator (ASI) corrected only for instrument error.

Calibrated airspeed (CAS) is the IAS corrected for instrument and installation errors. Equivalent air speed (EAS) is the IAS corrected for pressure and compressibility errors. True air speed (TAS) is the EAS corrected for density errors. The critical correction factor is derived from an outside air temperature (OAT) reading. With knowledge of OAT, these EAS correction factors can be derived from flight test data for given flight conditions, or automatically via software in computerized equipment and systems. TAS is a very important value needed for navigation purposes. If an aircraft is flying at 400 knots TAS in still air, its speed over the ground is 400 knots. If it now encounters a head wind of 50 knots, the speed over the ground is reduced to 350 knots.

> **KEY POINT**
>
> For a standard day, at mean sea level, with no position or instrument errors: IAS=CAS=EAS=TAS.

To illustrate the relationship between these terms, consider an IAS of 280 knots, referring to Table 2.2 (altitude in feet, speeds in knots).

Table 2.2 Relationship between altitude and aircraft speed

Altitude	IAS	CAS	EAS	TAS
10,000	280	280	277	323
20,000	280	280	273	375
30,000	280	280	268	437

KEY POINT

Indicated airspeed is the speed of an aircraft as shown on its pitot static-airspeed indicator calibrated to reflect standard atmosphere adiabatic compressible flow at sea level uncorrected for airspeed system errors.

KEY POINT

Calibrated airspeed is indicated airspeed of an aircraft, corrected for position and instrument error. Calibrated airspeed is equal to true airspeed in standard atmosphere at sea level.

KEY POINT

Equivalent airspeed is the calibrated airspeed of an aircraft corrected for adiabatic compressible flow for the particular altitude. Equivalent airspeed is equal to calibrated airspeed in standard atmosphere at sea level.

KEY POINT

True airspeed means the airspeed of an aircraft relative to undisturbed air.

ASI maintenance

ASIs normally do not require a great deal of maintenance when installed, other than visual inspection and a zero-reading check. When installing a new or repaired/overhauled instrument, it will have been fully checked by the manufacturer or repair shop. Typical on-aircraft checks include checking that the indicator reads zero when the aircraft is not moving and leak/range tests; details will be given in the aircraft maintenance manual.

KEY POINT

When testing an ASI through simulated range of air speeds, do not drive the indicator beyond its limits, this could damage the instrument.

MACHMETERS

Overview

At high air speeds, air loads on the airframe have an increasing effect. During flight, the aircraft disturbs the surrounding air and this sets up pressure waves – see Figure 2.11. The pressure waves move away from the aircraft at the speed of sound. At relatively low speeds the pressure waves move well ahead of the aircraft; the airflow around the aircraft remains reasonably smooth. As the aircraft approaches the speed of sound, it closes in on its own pressure wave and the air starts to compress on wing-leading edges and become turbulent around the airframe. These changes have an adverse effect on lift, drag, stability and aircraft controls. (For more on high-speed flight, see a new title in this series, *Aircraft Aerodynamics, Flight Control and Airframe Structures*.)

For all aircraft types capable of high-speed flight, a critical speed is established, above which the effects of compressibility have adverse effects. The aircraft speed itself is not the critical factor; it is the ratio of the aircraft's true airspeed to the local speed of sound that is important at high speeds – see Figure 2.12. The speed at which sound waves travel in a medium is dependent on the temperature and density of the surrounding air. The denser the air, the faster is the speed of the sound. For an aircraft in flight, the Mach number (named after the Austrian physicist Ernst Mach, 1838–1916) is defined as the ratio of the aircraft's true airspeed divided by the local speed of sound. From an aircraft instrumentation and performance perspective, the speed of sound reduces with altitude, as illustrated in Figure 2.13, and remains constant above the tropopause.

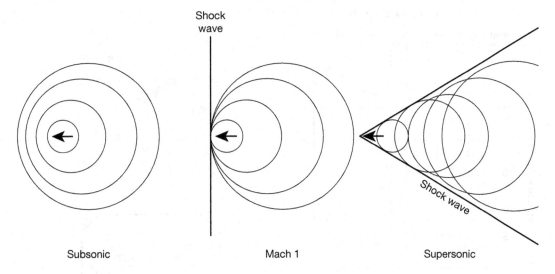

2.11 Pressure waves

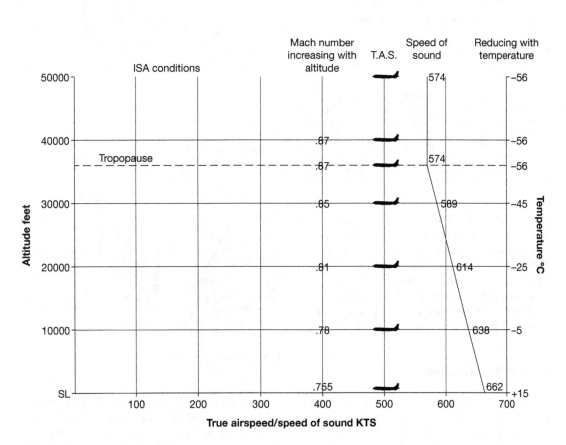

2.12 Mach number with altitude/ 1

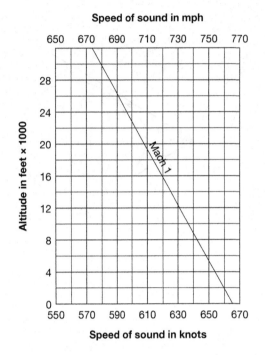

2.13 Mach number with altitude/ 2

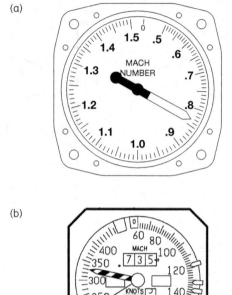

(a)

(b)

2.14 (a) Machmeter display (b) Mach/ASI display

The aircraft's critical Mach number is expressed as Maximum Mach Operating (MMO); this allows a margin below MMO for normal manoeuvres and turbulence, and so on. Some aircraft are installed with a dedicated instrument to indicate Mach number; some aircraft have a combined Mach/ ASI function – see Figure 2.14. For illustration purposes, a (simplified) mechanical instrument is described.

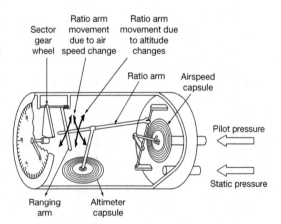

2.15 Machmeter schematic

Machmeter principles

The equations for true air speed and speed of sound both incorporate air density. In deriving the equation for Mach number, density is cancelled out and so the equation for Mach number is derived from dynamic and static pressures. In a Machmeter, the ratio of true air speed and local speed of sound is derived from an air-speed mechanism corrected by an altitude mechanism. Referring to the Machmeter schematic

(Figure 2.15), the airspeed capsule is connected directly to pitot pressure and is exposed externally (within the case) to static pressure. The air-speed capsule therefore responds to the difference between pitot–static pressure. As the capsule expands with increasing air speed, it causes the ratio arm to move in the direction shown. This movement is transmitted through the ranging arm and sector gear wheel to cause the pointer to move over the scale.

Machmeter schematic

The evacuated altitude capsule is exposed to static pressure within the case, thereby only responding to altitude changes. Altitude capsule movement is transmitted through the ratio arm as shown. The result of these two combined movements is transmitted through the ranging arm and sector gear, causing the pointer to move over the calibrated dial.

Machmeter maintenance

Machmeters normally do not require a great deal of maintenance when installed, other than visual inspection and a low speed-reading check (the pointer should be on the lower stop with no air speed). When installing a new or repaired/overhauled instrument, it will have been fully checked by the manufacturer or repair shop. Typical on-aircraft checks include leak/range tests; details will be given in the aircraft maintenance manual.

> **KEY POINT**
>
> When testing a Machmeter through a simulated range of air speeds, do not drive the indicator beyond its limits – this could damage the instrument.

PITOT STATIC SYSTEMS

Introduction

The four air-data instruments described in this chapter all require atmospheric pressure inputs:

Altimeter – static pressure
Vertical-speed indicator – static pressure
Air-speed indicator – static and pitot pressure
Machmeter – static and pitot pressure

Not all aircraft types are fitted with all four instruments; typically, smaller general-aviation aircraft have three instruments all connected to a pitot-static system, as shown schematically in Figure 2.16(a). High-speed aircraft include a Machmeter in the system, Figure 2.16(b). Larger transport aircraft have duplicated/balanced pitot-static systems, as illustrated in Figure 2.17. The physical arrangement of any pitot-static system is carefully designed to ensure high reliability of each instrument. The maintenance of pitot-static systems includes various practical tasks and considerations for safe operation.

There are three pressures associated with aircraft instruments: static, dynamic and pitot. Static pressure is exerted by still air, acting equally on all parts of the aircraft airframe. Dynamic pressure results from the aircraft moving through the air; it is proportional to the forward speed and density of the surrounding air. Pitot pressure is air that enters into the sensing head, or probe. It is the sum of static and dynamic pressures.

Pressure heads

There are two types of pressure head. The first type measures pitot pressure only, referred to as a pitot tube (Figure 2.18); the second type is a static tube; the third type is a combined pitot-static sensor. Either type can be installed either on the nose of the aircraft, protruding into the airflow ahead of the aircraft, on the side of the aircraft, or on the wing. Depending on the type certification of the aircraft, the pressure head might be heated to prevent ice formation. This is achieved by a system of heating elements inside the housing. These elements can be installed in the tube section alone and also in the support mast. The heating elements are powered from the aircraft's electrical system, controlled by the pilot.

> **KEY POINT**
>
> Only operate the pitot-head heating in accordance with the maintenance manual

(a)

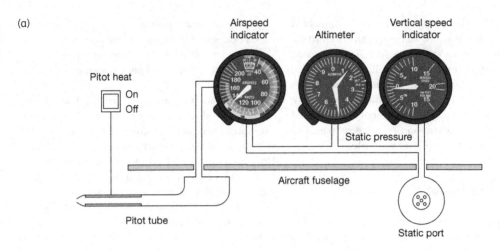

(b)

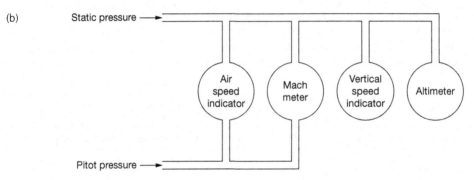

2.16 (a) Pitot-static system schematic (b) Pitot-static system schematic (with Machmeter)

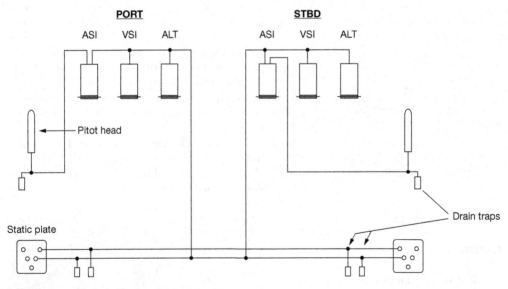

2.17 Duplicated/balanced pitot-static system

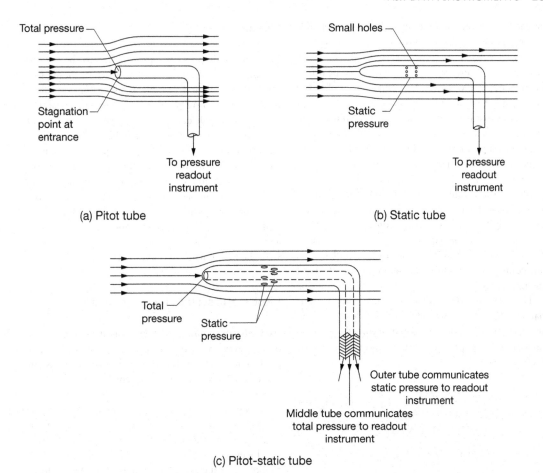

2.18 Pitot tube/static tube/pitot-static head

Pitot-head heating elements get very hot and can cause damage to equipment and are a personal safety hazard. This is particularly prevalent in still-air conditions where there is no cooling airflow.

Drain holes are located at the base of pressure heads to prevent the accumulation of moisture in the head. It is very important that these holes do not become blocked or damaged. Some pressure heads are designed for the port or starboard side of the aircraft and can be 'handed'. Pressure heads should not be painted since this could affect the thermal efficiency or block the static ports or drain holes.

Static vents

Static vents, or ports, can be incorporated into the pitot head as previously described, or they can be designed as separate items. The typical static vent (or port) installation is a flat metal plate secured to the fuselage skin – see Figure 2.19. The choice of integrated pitot/static heads or separate vents depends on a number of factors, including position error and ice

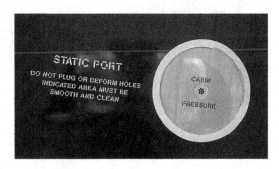

2.19 Static port

formation. The physical arrangement varies depending on the aircraft manufacturer. A simple system comprises a flat brass plate with a small (typically quarter-inch) diameter hole in its centre; a pipe fitting is connected inside the fuselage and this supplies the static supply to each of the instruments.

Aircraft that have several independent pitot-static systems typically have static vents incorporated into a stainless-steel plate. Two or more small holes are connected to static pipes for a specific system, e.g. left and right. The pipe is then connected via a threaded coupler. In this arrangement, static vents are fitted on both sides of the aircraft. This provides some redundancy – e.g. if one side were to become damaged from a bird strike, then the opposite side will still function. Having static vents on both sides of the fuselage and interconnected pipes also balances out the effects of yawing/side-slip.

The static vent outer plate(s) are fitted in positions that will have been evaluated during design of the aircraft to be free from the effects of turbulence from other aircraft equipment and structure, e.g. antennas, rivets and so on.

Pipelines

Pitot and static pressures are supplied to the various instruments through the respective pipes; these pipes are normally made from quarter-inch (or equivalent metric dimension) light alloy or tungum material. The pipes are normally identified with a label – 'Instrument air' or something similar. It is vital that the pipes remain clear of moisture; this can block the pipe as a liquid or ice. In addition to moisture drainage at the pressure-sensing heads and static vents, there are additional means of trapping and clearing moisture in the pipeline. Typical traps and drains are located at low points in the system, as illustrated in Figure 2.20. These traps and drains are cleared on a periodic basis, in accordance with the maintenance schedule/manual.

Pitot static system maintenance

As with any maintenance statements made in this book, always refer to the approved aircraft data and applicable safety instructions; the following is for training/educational purposes only.

- When removing pipes/hoses/equipment from the system, blanks are normally fitted to connections to prevent the ingress of moisture or foreign objects.
- Before installation, pipelines should be blown through with clean, dry, low-pressure air to clear any obstructions and/or moisture.

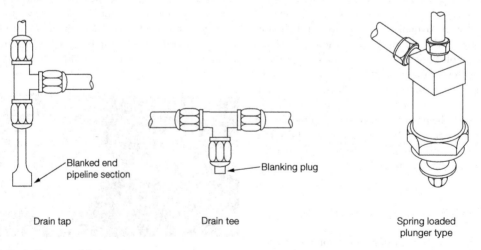

Drain tap · Blanked end pipeline section

Drain tee · Blanking plug

Spring loaded plunger type

2.20 Typical traps and drains

- When connecting flexible hoses, do not over-tighten or twist the hose.
- Instruments that require both pitot and static pressure, e.g. the ASI, normally have different size connections to prevent incorrect fitting.
- Leak-test the system after any removal/installation/disturbance.
- When leak testing, apply and release pitot/static pressures slowly to avoid damage to instrument capsules/diaphragms.
- When the aircraft is parked, cover pitot heads and plug static vents with the approved items to prevent moisture/foreign-object ingress. These items must be removed before flight.

ALTITUDE AND AIRSPEED WARNINGS/ALERTS

There are a number of miscellaneous airspeed and altitude warnings used on aircraft depending on

TEST YOUR UNDERSTANDING 2.5

What would happen to an altimeter and VSI if pitot pressure was inadvertently connected to the static connection?

TEST YOUR UNDERSTANDING 2.6

What would happen to an ASI, VSI and altimeter if there was a leak in the static pipe?

their type, certificate basis and/or operational use. These can either be stand-alone switches or they can be incorporated into other equipment. Typical functionality of these switches is illustrated in Figure

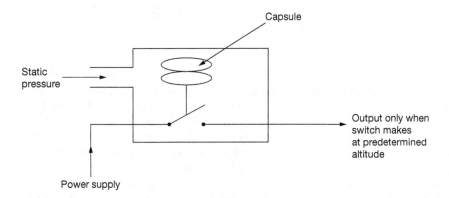

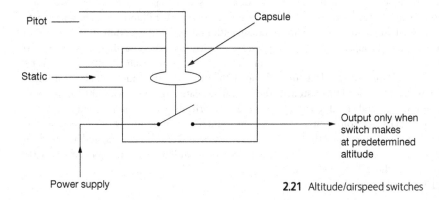

2.21 Altitude/airspeed switches

2.22 Altitude alert control unit

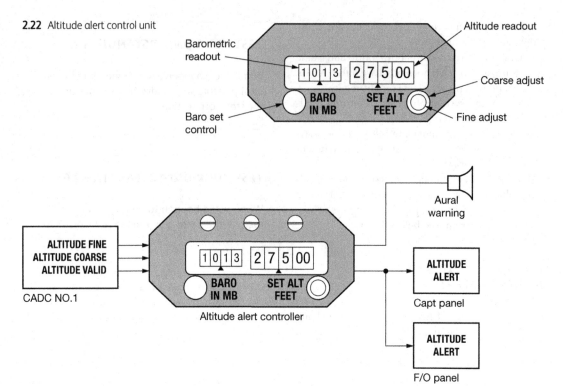

2.23 Altitude alerter interfaces

2.21. Airspeed and altitude switches are often used in the following systems:

- Air conditioning
- Cabin altitude
- Mach trim
- Configuration settings (e.g. flaps, landing gear, throttle setting)

Pressure switches can be designed with contacts that open or close at a preset altitude or airspeed.

These systems are described in more detail in another title in this book series, *Aircraft Electrical and Electronic Systems* (AEES).

Some aircraft are installed with altitude alert (or altitude preselect) systems. The purpose of this system is to provide the pilot with an aural and visual warning of either approaching or deviating from a pre-set altitude. The alert is given at a predetermined threshold, typically 1,000 feet.

Referring to Figure 2.22, the referenced barometric value is selected in accordance with the desired setting, e.g. QNE, QNH or QFE. The desired altitude is selected via coarse- and fine-control knobs. The controller is interfaced to an altitude source, e.g. an altimeter or air-data computer; these interfaces are illustrated in Figure 2.23. The physical interface will vary depending on aircraft and equipment type; typical interfaces are via synchros or a data bus.

The altitude-alert system operates as illustrated in Figure 2.24. As the aircraft climbs to within 900 feet of the preselected altitude a warning sound activates and an annunciator illuminates. The annunciator remains illuminated until within 300 feet of the preselected altitude; when passing through the 300 feet threshold the annunciator switches off. When flying at the preselected altitude, all alerts are off; if the aircraft deviates by more than 300 feet from the preselected altitude the warning sound is activated and the annunciator illuminates. If the aircraft continues to deviate from the preselected altitude, the alerts remain on until the deviation goes beyond 1,000 feet.

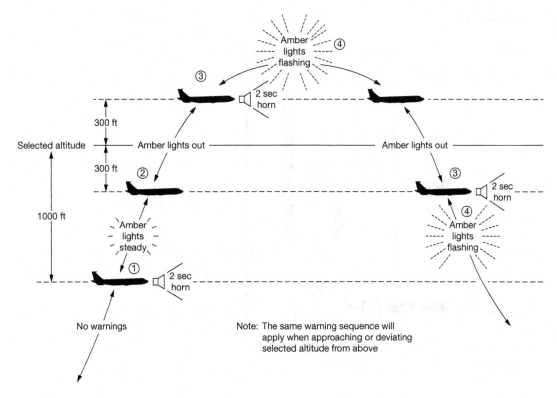

2.24 Altitude alert system operation

AIR-DATA COMPUTERS

The preceding sections of this chapter have described individual air-data instruments and alerting systems. Some aircraft are installed with equipment that incorporates one or more of the air-data functions, e.g. Mach number, true air speed, indicated air speed and so on. To compute any of these functions, pitot and static pressure inputs are required together with outside air temperature. A typical block diagram for an air-data computer (ADC) is shown in Figure 2.25. Air-data computers can either be incorporated into panel-mounted displays, or remotely located in the equipment bay. There is a wide range of technology used for the design and construction of air-data computers, including electromechanical and digital systems.

Air-data computers can be installed to serve as a centralized source of air-data (central air-data computer, CADC) information for use by other systems, e.g. flight control systems. (This topic is covered in

more detail in subsequent chapters.) Modern aircraft systems feature a high degree of integration; the Aspen Evolution Flight Display (EFD) incorporates all pitot-static functions within the panel-mounted electronic display – Figure 2.26. (Also covered in more detail in subsequent chapters.) This integrated technology replaces mechanical pressure-sensing capsules with solid-state micro-electrical-mechanical sensors (MEMS). MEMS is a technology that can be defined as miniaturized electromechanical elements that are made using the techniques of micro-fabrication. The physical dimensions of MEMS devices vary from one micron to several millimetres. The types of MEMS devices can vary from relatively simple structures having no moving elements to extremely complex electromechanical systems with multiple moving elements, all under the control of integrated microelectronics. There are at least some elements having some sort of mechanical functionality, whether or not these elements can actually move.

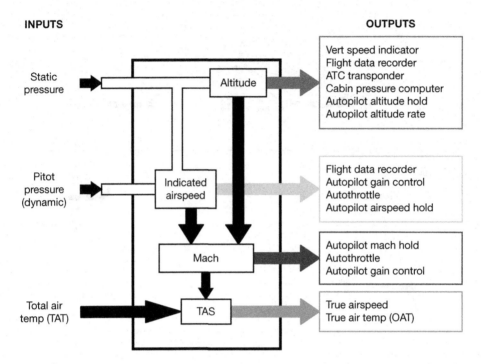

2.25 Air-data computer (ADC) block diagram

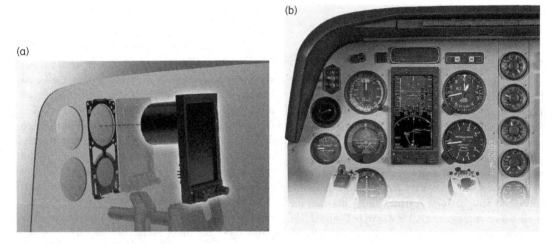

2.26 Evolution Flight Display (EFD): (a) Retrofit principles (b) Cockpit layout (single, dual and triple displays)

REDUCED VERTICAL SEPARATION MINIMUMS (RVSM)

RVSM provides access to previously restricted flight levels in upper airspace. With approval from the national aviation authority/EASA, RVSM allows air-craft to fly with a vertical separation of 1,000 feet between FL290 and FL410 inclusive. (This is reduced from 2,000 feet applicable to non-RVSM approved aircraft.)

To gain RVSM approval, the aircraft operator has to provide evidence that:

1. The RVSM airworthiness approval has been obtained
2. Procedures for monitoring and reporting height-keeping errors have been established
3. A training programme for the flight crew involved in these operations has been established
4. Operating procedures have been established.

Item 4 includes the avionic equipment to be carried, including its operating limitations and appropriate entries in the minimum equipment list (MEL). In addition to the equipment required by other certification and operating rules, aircraft used for operations in RVSM airspace have to be equipped with:

1. Two independent altitude-measurement systems
2. An altitude alerting system
3. An automatic altitude-control system
4. A secondary surveillance radar (SSR) transponder with altitude-reporting system that can be connected to the altitude-measurement system in use for altitude control.

Operators have to report recorded or communicated occurrences of height-keeping errors caused by malfunction of aircraft equipment or of operational nature, equal to or greater than:

*   Total vertical errors (TVE) of ±90 m (±300 ft);
*   Altimetry system errors (ASE) of ±75 m (±245 ft); and
*   Assigned altitude deviations (AAD) of ±90 m (±300 ft).

ADVANCED SENSORS

Pitot heads on new aircraft are being installed with integrated sensors known as SmartProbes® that contain an internal air-data computing function to calculate air-data parameters – all pneumatic tubing and considerable electrical wiring and other components are eliminated. These advanced probes integrate pressure sensors and air-data computer processing to provide all critical air-data parameters, including pitot and static pressure, air speed, altitude, angle of attack and angle of sideslip. The SmartProbe® system interfaces with total air temperature to provide static air temperature and true air speed. The system renders greater accuracy from the air-data output

for improved performance and reduced life-cycle costs.

MULTIPLE-CHOICE QUESTIONS

1. The altimeter's capsule is:

    (a) Connected to atmospheric pressure
    (b) Evacuated and sealed
    (c) Connected via a calibrated orifice

2. The critical Mach number of an aircraft determines:

    (a) The maximum speed the aircraft can fly
    (b) The minimum speed the aircraft can fly
    (c) The local speed of sound

3. When QNH is set on the altimeter, the displayed units are referred to as:

    (a) Altitude
    (b) Height
    (c) Pressure altitude expressed in flight levels

4. Port and starboard static vents are interconnected to minimize errors caused by:

    (a) Leaks
    (b) Compressibility
    (c) When the aircraft yaws

5. If the atmospheric pressure is 1,020mb and the altimeter's baro-scale was set to 1013.25mb, the altimeter should read:

    (a) Above zero feet
    (b) Below zero feet
    (c) Zero feet

6. RVSM allows aircraft to fly with a vertical separation of:

    (a) 2,000 feet between FL290 and FL410 inclusive
    (b) 1,000 feet between FL290 and FL410 inclusive
    (c) 1,000 feet up to FL290

7. A pitot-static leak test is carried out:

   (a) Whenever the system is disturbed
   (b) Only when a leak is reported
   (c) Only when an ASI is changed

8. If an aircraft is flown at a constant true air speed of 200 knots during a climb from sea level to 40,000 feet, the Machmeter will indicate:

   (a) Decreasing Mach number
   (b) Constant Mach number
   (c) Increasing Mach number

9. In level flight, a blockage of the VSI-metering unit would indicate the aircraft:

   (a) Having zero rate of climb
   (b) Climbing
   (c) Descending

10. Dynamic pressure sensed by an ASI varies with the:

   (a) Square root of speed
   (b) Square of speed
   (c) Rate of climb

3   Gyroscopic instruments

Gyroscopic instruments provide the pilot with additional displays and information needed for the basic operation of an aircraft, including attitude and direction. This chapter describes the various instruments used to achieve these displays, together with the additional data that can be derived. Before delving into the various gyroscopic instruments and systems, some generic principles of gyroscopes are given. The type and quantity of gyroscopic instruments and systems installed on an aircraft depends on a number of factors, ranging from aircraft size, type of operation, regulatory requirements and so on.

The chapter describes gyroscopic principles and the various instruments and systems installed on a range of aircraft, from small general aviation through to larger public-transport aircraft. Technology used across this range of gyroscopic instruments and systems ranges from simple electromechanical devices through to solid-state sensors and displays.

GYROSCOPE PRINCIPLES

Gyroscopic motion

Typical gyroscopic instruments are the artificial horizon and the turn coordinator, as shown in Figure 3.1. Gyroscopic principles are based on the physics associated with angular motion and the inertia and momentum of a body in circular motion. From Newton's laws, momentum is defined as the product of the mass of a body and its velocity. Momentum is a measure of the quantity of motion of a body. The force

Air speed indicator
Attitude indicator (artificial horizon)
Altimeter
Turn coordinator
Vertical speed indicator
Heating indicator

3.1 Primary flight instruments

that resists a change in momentum (i.e. resists acceleration) is known as inertia. A free gyroscope, as in Figure 3.2, is a rotating mass that has freedom to move at right angles to its plane of rotation. The rotor of a free gyroscope is mounted in a frame, or gimbol, such that it is free to turn in all directions and adopt any attitude. In this arrangement, the gyroscope has three degrees of movement:

1.   Spinning freedom; it is free to rotate on its own axis
2.   Veering freedom; it is free to turn about a vertical axis
3.   Tilting freedom; is free to turn about a horizontal axis.

Gyroscopic instruments utilize two fundamental characteristics of a gyro rotor, that of rigidity (or gyroscopic inertia) and precession.

Rigidity is an application of Newton's first law of

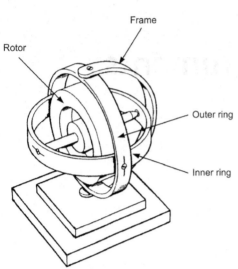

3.2 Free gyroscope

3.3 Sperry's rule of precession

motion, where a body remains in its state of rest or uniform motion unless compelled by some external force to change that state. If a gyro rotor is revolving it will continue to rotate about that axis unless a force is applied to alter the axis. The greater the momentum of the rotor, i.e. the heavier it is and the faster it rotates, the greater is the gyro's resistance to change, and so it has greater rigidity or inertia. The property of rigidity is used as the basis of providing a reference point in space under particular circumstances, no matter what the attitude of the aircraft. Precession may be defined simply as the reaction to a force applied to the axis of a rotating assembly.

KEY POINT

A gyroscopic rotor has rigidity and precesses when acted upon by an external force applied to the rotor assembly.

Laws of gyro-dynamics

The two properties of rigidity and precession provide the visible effects of the laws of gyrodynamics, which may be stated as follows:

1. If a rotating body is mounted so as to be free to move about any axis through the centre of mass, then its spin axis remains fixed in inertial space no matter how much the frame may be displaced.
2. If a constant torque is applied about an axis, perpendicular to the axis of spin, of an unconstrained, symmetrical, spinning mass, then the spin axis will precess steadily about an axis mutually perpendicular to both spin and torque axis.

Sperry's rule of precession

Elmer Ambrose Sperry, Sr. (1860–1930) was an American inventor and entrepreneur who, along with the German inventor Hermann Anschütz-Kaempfemost (1872–1931), is associated with the development and manufacture of compasses and stabilizers adopted across industry for a variety of applications, including aircraft autopilots.

The direction in which precession takes place is dependent upon the direction of rotation for the mass and the axis about which the torque is applied. Sperry's rule of precession, illustrated in Figure 3.3, provides a guide to the direction of precession, knowing the direction of the applied torque and the direction of rotation of the gyro-wheel.

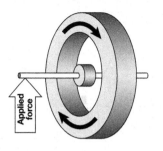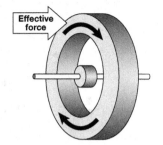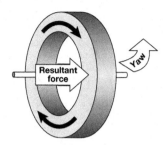

3.4 Gyroscopic precession

If the applied torque is created by force acting at the inner gimbol, perpendicular to the spin axis, it can be transferred as a force to the edge of the rotor, at right angles to the plane of rotation. The point of application of the force should then be carried through 90 degrees in the direction of rotation of the mass and this will be the point at which the force appears to act. It will move that part of the rotor rim, in the direction of the applied disturbing force; rotor precession is illustrated in Figure 3.4.

Gyroscopic wander

Movement between the spin axis and its frame of reference is called gyroscopic wander; referring to Figure 3.5, this has two main causes. Real wander is the actual misalignment of the spin axis due to mechanical defects in the gyroscope. Apparent wander is the discernable movement of the spin axis due to the reference frame in space, rather than spin axis misalignment. Wander in a gyroscope is termed drift or topple, dependent upon the axis about which it takes place. If the spin axis wanders in the azimuth plane it is known as drift and in the vertical plane it is referred to as topple.

Thus in real wander, the problems of friction in the gimbol bearings and imperfect balancing of the rotor cause torques to be set up perpendicular to the rotor spin axis; this leads to precession and actual movement, or real wander of the spin axis. There are two main causes of apparent wander, one due to rotation of the earth and the other due to movement over the earth's surface of the aircraft, carrying the gyroscope.

Figure 3.6 illustrates the various effects of apparent wander. With the gyro at the North Pole and the axis of spin vertical, (a), the spin axis and earth's axis coincide, there is no apparent wander of the gyro due to the earth's rotation. If the gyro were located at the equator (b), with spin axis initially vertical at say noon, it will appear to rotate in a vertical plane such that six hours later the spin axis will become vertical with respect to the earth's surface (even though the gyro spin axis has not changed). This apparent movement of the spin axis continues through the 24-hour rotation of the earth.

> ### TEST YOUR UNDERSTANDING 3.1
>
> What is the maximum rate of apparent gyro wander at the equator?

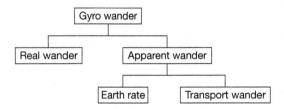

3.5 Gyroscopic wander

With a gyroscope located at the North Pole, and spin axis horizontal (c), the earth will rotate and, to an observer, the spin axis will appear to wander at 15 degrees per hour in azimuth in the opposite direction. Finally, with the gyro located at the equator (d), with spin axis horizontal pointing N-S, the spin axis is now parallel with the earth's axis, and will continue in this direction; there is no apparent wander in this situation.

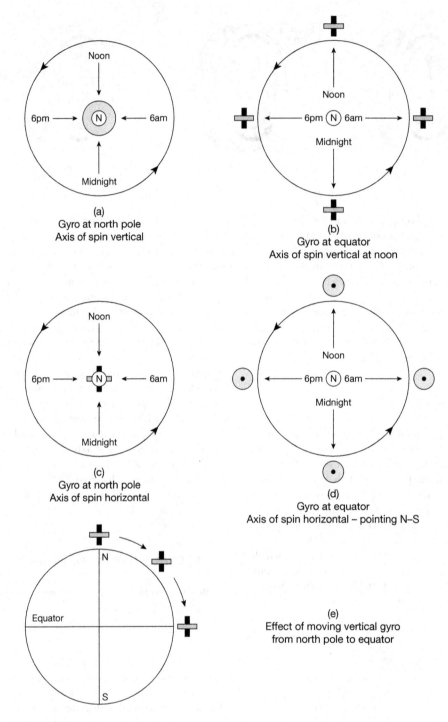

(a)
Gyro at north pole
Axis of spin vertical

(b)
Gyro at equator
Axis of spin vertical at noon

(c)
Gyro at north pole
Axis of spin horizontal

(d)
Gyro at equator
Axis of spin horizontal – pointing N–S

(e)
Effect of moving vertical gyro
from north pole to equator

3.6 Apparent wander

The above scenarios have considered the effects of specific gyroscope locations and the earth's rotation; a similar effect occurs if the gyroscope is moved over the earth's surface (e). Moving the gyroscope from the North Pole to the equator leads to transport wander.

PRACTICAL GYROSCOPES

The spinning rotor of a gyroscope can either be driven by airflow or electrical energy. The basic principle and laws of the gyro still apply, but the means of implementing the design vary. Air driven gyroscopic instruments are still used in many general aviation (GA) applications. (In the general aviation (GA) market, electrically driven gyroscopic instruments tend to be more expensive.)

An electrically driven gyro operates as an electric motor, with the spinning wheel acting as the motor armature. The motor can either be driven from a 14/28 DC or 26 V 400 Hz AC supply depending on the aircraft type. Typical gyro wheel speed for a DC-driven rotors is 8,000 rpm; for an AC-driven rotor at 400 Hz will be a nominal 24,000 rpm.

KEY POINT

Low gyro wheel speeds cause slow instrument response or lagging indications; fast gyro wheel speeds cause the instruments to overreact.

Utilizing the gyro's freedom of movement, stability and known properties, various instruments are used in the aircraft to provide reference information to the pilot.

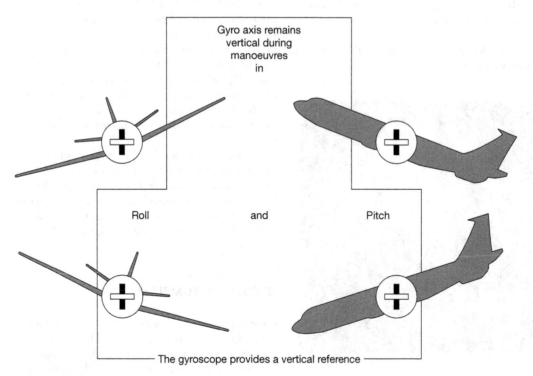

3.7 Gyro as a vertical reference

ARTIFICIAL HORIZONS

The artificial horizon, or attitude indicator, is based on a vertical gyro (VG), where the gyro's spin axis is maintained in the vertical direction relative to the earth's surface – see Figure 3.7. In this application, the gyroscope provides a vertical reference to the pilot. Referring to basic gyroscope theory, the VG will exhibit characteristics associated with real and apparent wander. The gyro's rotor has to be corrected for these two effects; this is achieved by sensing the gravitational acceleration of the earth and applying torque to the gimbals causing the gyro to precess into the required position. The freedom to turn about a vertical and horizontal axis is based on the displacement gyroscope; pitch and roll manoeuvres can now be displayed by the amount of displacement of the gimbals.

The terms artificial horizon and attitude indicator often get interchanged. Historically, the artificial horizon was an instrument that only displayed vertical reference information, whereas the attitude indicator incorporates additional features, e.g. electrical signals to other systems including weather radars and autopilots. Referring to Figure 3.8, the typical external features of an attitude indicator are:

• The 'model aircraft' representing the wings and nose of the aircraft

3.8 Attitude indicator

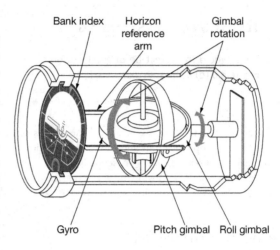

3.9 Typical internal features of an attitude indicator

• The 'horizon' separating the two halves of the display
• Graduating marks representing the roll/bank angle.

The upper half of the display, above the horizon, is usually blue in colour to represent sky; the lower half of the display, below the horizon, is usually light brown to represent the earth. Note that the horizon moves up and down and banks, while the aircraft symbol is fixed relative to the rest of the instrument panel. Typical internal features shown in Figure 3.9 include the gyroscope, gimbals and bank-angle display. Artificial horizon, or attitude indicator, displays during typical pitch and roll manoeuvres are shown in Figure 3.10.

Another development of the attitude indicator is the attitude direction indicator (ADI) – this is typically installed on larger aircraft. The ADI can either be an electromechanical instrument or an electronic ADI (EADI) – see Figure 3.11. Additional features of an ADI include ILS, flight director and auto-throttle displays.

GYROSCOPE TUMBLING

If the applied forces to a gyroscope are too hard or rapid, the gyro can become unstable rather than precess, a condition known as tumbling, or toppling.

This is a phenomenon in which the gyro axes precess at random, giving erratic indications caused

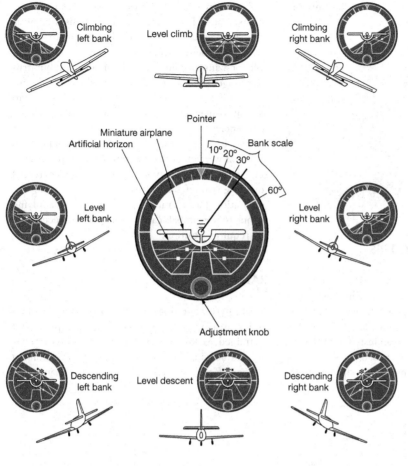

3.10 Typical manoeuvres and the attitude indicator

Climbing left bank

Level climb

Climbing right bank

Pointer

Miniature airplane
Artificial horizon

Bank scale

10° 20°
30°

60°

Level left bank

Level right bank

Adjustment knob

Descending left bank

Level descent

Descending right bank

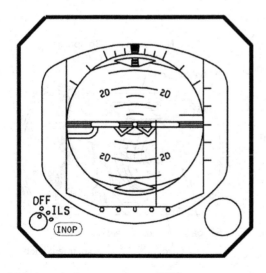

3.11 Attitude direction indicators (ADI)

by the gyro gimbals coming against their mechanical stops and locking up. The gyro must be re-erected in a straight and level flight before the instrument can give correct indications. Most artificial horizons have a 'pull to cage' knob that aligns the gimbals back to their neutral position.

As previously described, precession is caused by aircraft manoeuvring and by the internal friction of attitude and directional gyros. This precession causes slow drifting and thus erroneous readings. Tumbling a gyro should be avoided since it damages bearings and renders the instrument unusable until the gyro is erected again. Some gyro instruments have a mechanical frame, or cage, that is temporarily engaged to hold the gimbals in place. Modern gyro instruments have higher attitude limitations; these do not tumble if the gyro limits are exceeded. To achieve this, they do not indicate pitch attitude beyond +/− 85 degrees nose up/down, and have self-erecting mechanisms that eliminate the need for caging.

DIRECTIONAL GYROSCOPES

The directional gyroscope (DG) is used to sense direction in the azimuth plane; it has a horizontal spin axis. A horizontal gyroscope's schematic and principles of operation are shown in Figure 3.12. The DG is normally incorporated into an instrument that has a compass card display; this is used for short-term heading stability during turns. In this application, the

DG is set by comparison to the aircraft's magnetic compass and checked on a periodic basis. Magnetic compasses are unreliable in the short term, i.e. during turning manoeuvres. Directional gyroscopes are reliable for azimuth guidance in the short term, but drift over longer time periods. A combined magnetic compass stabilized by a directional gyroscope (referred to as a gyro-magnetic compass) can overcome these deficiencies.

As with all gyroscopes, the DG will be subject to apparent wander, both earth rate and transport rate – see Figure 3.13. The gyro's rotor has to be corrected for these two effects; this is achieved by sensing the gravitational acceleration of the earth and applying torque to the gimbals causing the gyro to precess into the required position.

RATE GYROSCOPES

A rate gyroscope does not have the same freedom as the displacement gyroscope; it is captive and restrained, as illustrated in Figure 3.14. The rotor has one gimbal ring, held in position by two restraining springs. Rate gyroscopes respond to changes in

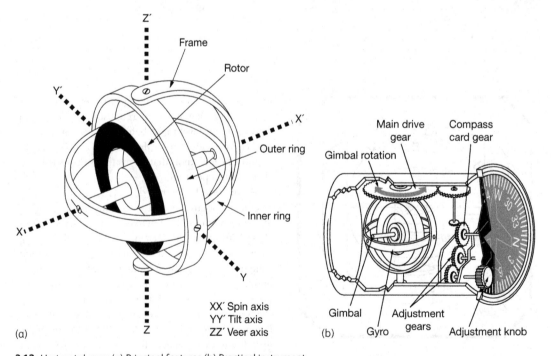

3.12 Horizontal gyro: (a) Principal features (b) Practical instrument

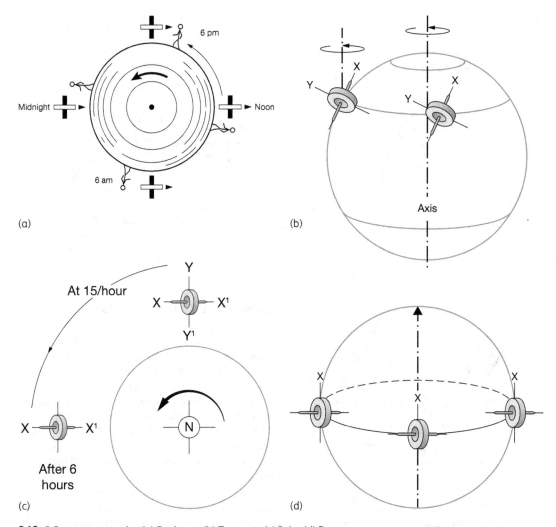

3.13 DG apparent wander: (a) Earth rate (b) Transport (c) Poles (d) Equator

direction rather than displacement or magnitude of direction. The rotor is pivoted in the gimbal, which is pivoted in a fixed supporting frame. Springs maintain the gyroscope in the neutral position. The primary axis in this illustration is the operating axis of the gyroscope; in this orientation it will measure turn rate. When the frame turns about the primary axis, a torque is applied to the rotor via the gimbal and rotor pivots. As the gimbal moves from the neutral position, the restraining springs go into compression/tension, setting up an opposing torque on the gimbal. Precession stops when the spring torque force is equal to the precession force. The new position of the gyro is directly proportional to the rate of turn about the primary axis.

> **KEY POINT**
>
> The amount of rate gyro displacement is proportional to the rate of movement, rotor speed and spring tension.

(a)

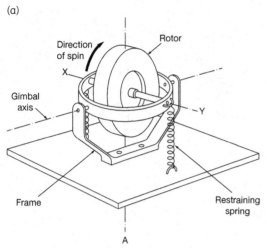

(b)

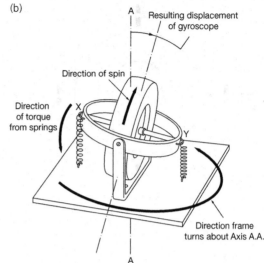

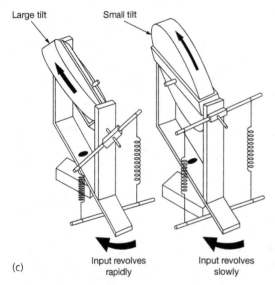

(c)

3.14 Rate gyroscope schematics: (a) Overview (b) Rate sensing (c) Varying rates change

TURN-AND-SLIP INDICATORS

Turn-and-slip indicators provide an indication of the rate of turn about the vertical axis of the aircraft and the amount of slip/skid arising from an incorrect (uncoordinated) turn. The turn rate is derived from a directional (horizontal) rate gyro – see Figure 3.15. A typical turn-and-slip indicator display is shown in Figure 3.16.

Referring to Figure 3.17, in level flight, the aircraft's lift is equal and opposite to the total lift (see Chapter 7). In a medium banked turn, total lift comprises two components, vertical and horizontal; bank angle loads now comprise the aircraft weight and centrifugal force, giving a resultant load. In a steep turn, these bank angle forces all increase. A given aircraft type will have a maximum bank angle and airspeed limitation to protect the aircraft structure from being over-stressed.

Slip/skid is displayed in the form of an inclinometer; slip and/or skid occurs when the aircraft is not turned in a coordinated turn manoeuvre – see Figure 3.18. This occurs when the aircraft's vertical axis deviates from the direction of gravity in straight flight, or from the resultant direction of gravity in a turn. During straight and level flight or during a coordinated turn, the aircraft's vertical axis aligns with the direction of gravity, or the resultant direction of gravity in a turn, and the ball remains in the centre of the inclinometer. During an uncoordinated turn, the aircraft's vertical axis does not align with the resultant direction of gravity, and the ball is offset in

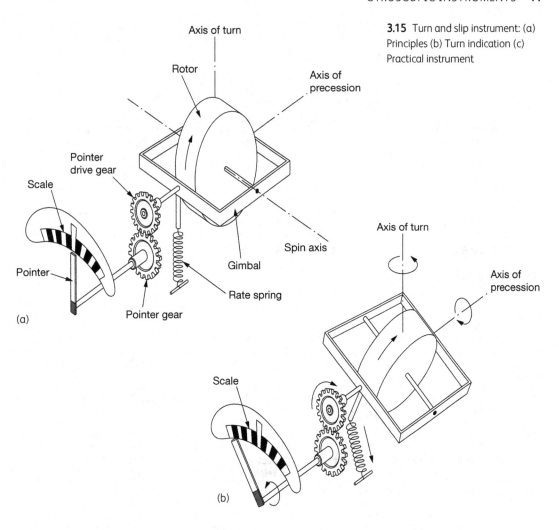

3.15 Turn and slip instrument: (a) Principles (b) Turn indication (c) Practical instrument

Axis of turn

Rotor

Axis of precession

Pointer drive gear

Scale

Spin axis

Pointer

Gimbal

Rate spring

Pointer gear

(a)

Axis of turn

Axis of precession

Scale

(b)

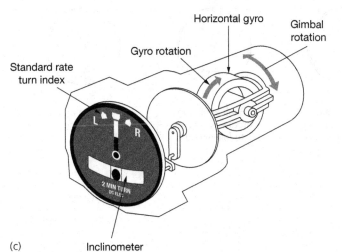

Horizontal gyro

Gimbal rotation

Gyro rotation

Standard rate turn index

L   R

2 MIN TURN
DC ELEC

(c)   Inclinometer

L   R

2 MIN TURN
D.C. ELEC

3.16 Turn and slip indicator

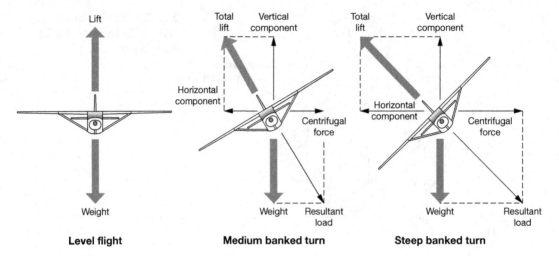

3.17 Bank angle loads

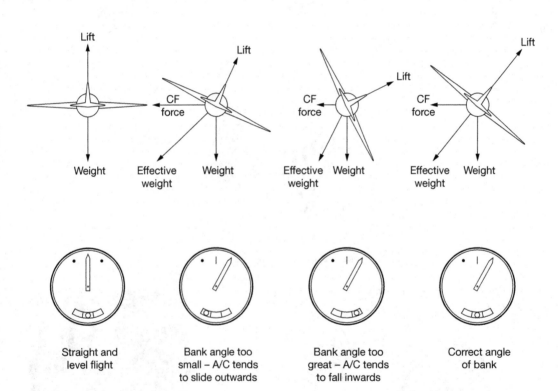

3.18 Coordinated turn manoeuvre

the inclinometer. The inclinometer is filled with fluid to dampen any transient movements.

TURN COORDINATORS

The turn coordinator is an alternative instrument, often used in place of the turn-and-slip indicator on small GA aircraft – see Figure 3.19. The primary difference is the calibration of the precession of the rate gyroscope – see Figure 3.20. This is set at a nominal 30 degrees with respect to the aircraft's longitudinal axis, thus making the gyroscope sensitive

3.20 Turn coordinator

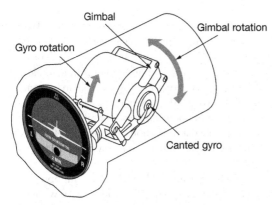

3.19 Turn coordinator schematic

to rolling and turning. Since a turn is initiated by banking, the gyroscope will precess, thereby moving the aircraft symbol to indicate the direction of the bank, enabling a pilot to anticipate the resulting turn. The turn is then controlled by the pilot to the required rate, as indicated by the alignment of the aircraft symbol with the graduations on the outer scale. Coordination of the turn is indicated by the inclinometer remaining centred – see Figure 3.21.

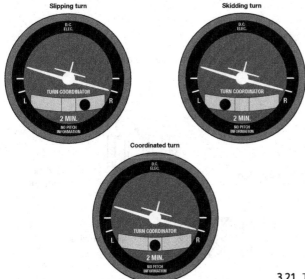

3.21 Turn coordinator displays

Air-driven gyros on GA aircraft are normally derived from a Venturi tube, or powered by a vacuum pump driven by the aircraft's engine – see Figure 3.22. Low pressure (3–5 ″Hg) pipelines connect the pump to the instruments, drawing air through filters in the instrument case. (Some GA aircraft have electrically driven vacuum pumps as a backup to the engine driven pump.) As the low-pressure air enters the instrument case, it is directed against small 'buckets' formed in the gyro's wheel. A regulator is attached between the pump and the gyro wheel case to control air pressure. From gyro principles, a constant wheel speed is essential for reliable instrument readings, and the correct air pressure is maintained with a pressure regulator. The typical air-driven gyro wheel speed is 15,000 rpm.

STRAPDOWN TECHNOLOGY

Introduction

The preceding sections of this chapter have described electromechanical gyroscopes. There are alternative 'strapdown' technologies available offering many advantages over electromechanical devices: no moving parts, increased reliability, and software-based functionality; the disadvantage is cost. This book does not attempt to make a case for either technology, but describes them for educational and training purposes.

Laser gyros

Ring laser gyros (RLG) use interference of a laser beam within an optic path, or ring, to detect rotational displacement – see Figure 3.23 – for measuring changes in pitch, roll and direction (azimuth). Note that laser gyros are not actually gyroscopes in the strict sense of the word – they are in fact sensors of angular rate of rotation about an axis. The primary use of RLGs is in inertial navigation systems – see Chapter 5 – and attitude heading reference systems (AHRS) – Chapter 6.

Two laser beams are transmitted in opposite directions (contra-rotating) around a cavity within a triangular block of cervit glass; mirrors are located

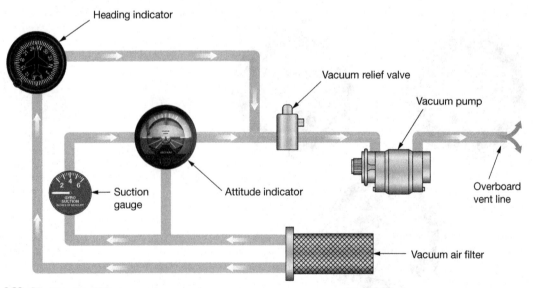

3.22 GA vacuum system

(a)

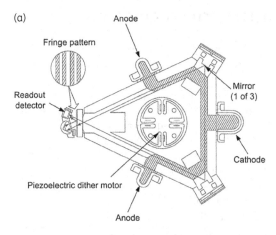

(b)

3.23: (a) Ring laser gyro schematic (b) Ring laser gyro

in two of the corners. The cervit glass (a ceramic material) is very hard and has an ultra low thermal expansion coefficient. The two laser beams travel the same distance but in opposite directions; with a stationary RLG, they arrive at the detector at the same time.

The principles of the laser gyro are based on the Sagnac effect, named after the French physicist Georges Sagnac (1869–1926). This phenomenon, encountered in interferometry, is caused by rotation. Interferometry is the science and technique of superposing (interfering) two or more waves, which creates a resultant wave different from the two input waves; this technique is used to detect the differences between input waves. In the aircraft RLG application, when the aircraft attitude changes, the RLG rotates; the laser beam in one path now travels a greater distance than the beam in the other path; this changes its phase at the detector with respect to the other beam. The angular position, i.e. direction and rate of the RLG, is measured by the phase difference of the two beams. This phase difference appears as a fringe pattern caused by the interference of the two wave patterns. The fringe pattern is in the form of light pulses that can be directly translated into a digital signal. Operating ranges of typical RLGs are 1,000 degrees per second in pitch, roll and azimuth. In theory, the RLG has no moving parts; in practice there is a device required to overcome a phenomenon called lock-in. This occurs when the frequency difference between the two beams is low (typically 1,000 Hz) and the two beams merge their frequencies. The

solution is to mechanically oscillate the RLG to minimize the amount of time in this lock-in region.

Ring laser gyros are very expensive to manufacture; they require very high quality glass, cavities machined to close tolerances and precision mirrors. There are also life issues associated with the technology. A variation of this laser gyro technology is the fibre-optic gyro (FOG), where the transmission paths are through coiled fibre optic cables packaged into a canister arrangement to sense pitch, roll and yaw – see Figure 3.24. The fibre-optic gyroscope also uses the interference of light through several kilometres of

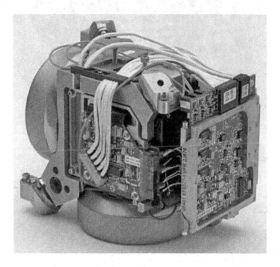

3.24 Fibre-optic gyro

coiled fibre-optic cable to detect angular rotation. Two light beams travel along the fibre in opposite directions and produce a phase shift due to the Sagnac effect. Fibre-optic gyros have a life expectancy in excess of 3.5 million hours.

MEMS technology

As with air data sensors, MEMS technology has enabled miniaturization and subsequent integration of micro-electronics. The Aspen EFD product, Figure 3.25, incorporates the equivalent functionality of vertical gyros, directional gyros and rate gyros. On a MEMS chip, movement of the silicon crystal structure causes a change in electrical current that can be measured and processed. Air data computer information and a GPS position help complete the attitude and reference information needed to accurately depict aircraft position in space. In Aspen's case a remote sensor module (RSM), which includes a magnetometer and a GPS sensor, completes the data needed for attitude/direction and rate calculations.

BUILT-IN
BACKUP
BATTERY

3.25 Aspen EFD

MULTIPLE-CHOICE QUESTIONS

1. The direction in which precession takes place in a gyroscope is dependent upon the:

   (a) Rotor mass and speed
   (b) Direction of rotation and the axis about which the torque is applied
   (c) Location on the earth's surface

2. The attitude indicator displays:

   (a) Pitch and roll
   (b) Heading
   (c) Rate of climb

3. The turn coordinator is often used in place of the:

   (a) Artificial horizon
   (b) Attitude indicator
   (c) Turn-and-slip indicator

4. Low gyro wheel speeds cause:

   (a) Slow instrument response or lagging indications
   (b) The instrument to overreact
   (c) No effect on the indication

5. Turn indicators are based on which type of gyroscope?

   (a) Displacement
   (b) Rate
   (c) Vertical

6. The turn coordinator displays:

   (a) Pitch attitude
   (b) Roll attitude
   (c) Pre-defined roll rate

7. The rate gyro responds to:

   (a) Changes in direction
   (b) Displacement
   (c) Magnitude of direction

8. The directional gyro (DG) is used to sense direction in the:

    (a) Azimuth plane, with a horizontal spin axis
    (b) Azimuth plane, with a vertical spin axis
    (c) Longitudinal plane, with a horizontal spin axis

9. Laser gyros sense

    (a) Heading
    (b) Angular rate of rotation about an axis
    (c) Displacement

10. Rigidity of a gyroscope increases with:

    (a) Applied torque
    (b) Location on the earth's surface
    (c) Rotor mass and speed

4 Flight instruments

Air data and gyroscopic instruments are referred to as the primary flight instruments. They are arranged in the instrument panel in one of several conventional ways – the basic 'flying tee' or its derivative, the 'six pack'. The instruments can contain their own sensors, e.g. in the case of an altimeter with internal capsule; alternatively, the sensor can be remote, e.g. a centralized air-data computer. Older aircraft that were installed with electromechanical instruments are often upgraded with modern electronic flight displays, featuring the same basic functions as electromechanical instruments, but with additional features. Electronic flight bags, although not flight instruments, are included in this chapter under the subject of electronic flight displays.

INSTRUMENT PANEL LAYOUT

The majority of aircraft built since the late 1940s and early 1950s have four flight instruments located in a standard arrangement called the flying T (or Tee) – see Figure 4.1. The artificial horizon/attitude indicator is in the top/centre; the air-speed indicator (ASI) to the left, altimeter to the right and directional gyro underneath. Developments in microprocessors and display technology have led to electronic versions of these instruments, often with additional features and benefits (these are described later in this chapter).

The turn-and-slip indicator (or alternatively a turn-coordinator) and vertical-speed indicator are usually located under the ASI and altimeter. The magnetic compass (Figure 4.2) will be located above

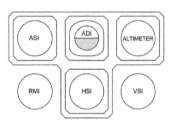

(a) Basic 'T' flight instrument configuration

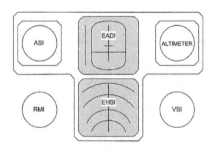

(b) Basic EFIS flight instrument configuration

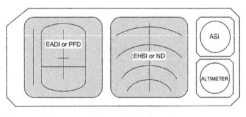

(c) Enhanced EFIS flight instrument configuration

4.1 'Flying T' arrangement (basic + electronic)

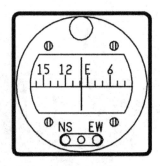

4.2 Magnetic compass display

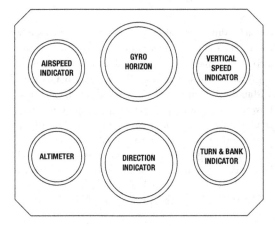

Basic 6 grouping

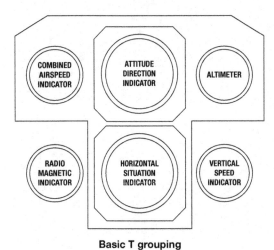

Basic T grouping

4.3 'Six pack' arrangement and flying tee

the instrument panel, normally on the windscreen centre post.

This instrument-panel arrangement was widely adopted throughout the UK military, from light single-engine trainers to four-engine heavy bombers, minimizing the type-conversion issues associated with 'blind flying'. This basic arrangement, also known as a 'six pack', was widely adopted by commercial aviation. In the late 1940s, the arrangement was changed, as illustrated in Figure 4.3.

4.4 Instrument panel: (a) GA aeroplane (b) Transport aeroplane

- Top row: airspeed, artificial horizon, altimeter
- Bottom row: radio compass, direction indicator, vertical speed.

For single-pilot aircraft, this instrument-panel arrangement would be directly in front of the pilot. Small general-aviation aeroplanes are traditionally based on the pilot in command being in the left-hand seat – Figure 4.4(a). On larger transport aeroplanes, the instruments are duplicated – see Figure 4.4(b). The pilot in command of a rotorcraft will be in the right-hand seat due to the arrangement of flying controls – Figure 4.5. Single-pilot aircraft have some of these primary instruments repeated on the opposite panel depending on the operational role of the aircraft, e.g. training aircraft will duplicate the primary instruments. Aircraft that are certified for two-crew operations will duplicate the primary instruments on both panels.

In good weather conditions (called visual flight rules – VFR) smaller GA aircraft can be operated through all flight phases without the use of any of the subject instruments. Aircraft approved to fly in instrument flight rules (IFR) do require these flight instruments. Larger two-crew aircraft are equipped with a third attitude indicator for standby and comparison purposes. The vertical-speed indicator is beneficial but not essential.

4.5 Rotorcraft instrument panel

can arise during the aircraft's manufacture, primarily the effect of the earth's field running through the airframe whilst on the production line and then (if applicable) whilst in storage. In addition, magnetism within the airframe can be induced through electrical power generation and avionics equipment, particularly radar. In some cases the aircraft will be degaussed, although some residual magnetism will remain. Throughout the life of the aircraft, there will be the need to carry out calibration of the on-board compass systems to check acceptable readings and adjust if necessary; this process is called a compass swing.

COMPASS INSTRUMENTS

Introduction

The earth is surrounded by a weak magnetic field, which can be represented by a bar magnet through the earth's centre; the resulting magnetic field converges at the north and south magnetic poles, as seen in Figure 4.6. Although the earth's magnetic field can be used to detect aircraft heading, there are practical considerations that have to be addressed, both for the earth's field itself and also the magnetic properties of the aircraft. As described in Chapter 5, true north and magnetic north are in different locations; the difference between true and magnetic north will vary at most places on earth by some amount, referred to as variation.

An aircraft contains both soft and hard ferrous-based materials, giving rise to permanent and temporary magnetism. Permanent magnetic effects

KEY POINT

The difference between true and magnetic north at any point on the earth's surface is called variation.

KEY POINT

Magnetic interference caused by the aircraft's structure and components is called deviation; its magnitude varies from aircraft to aircraft.

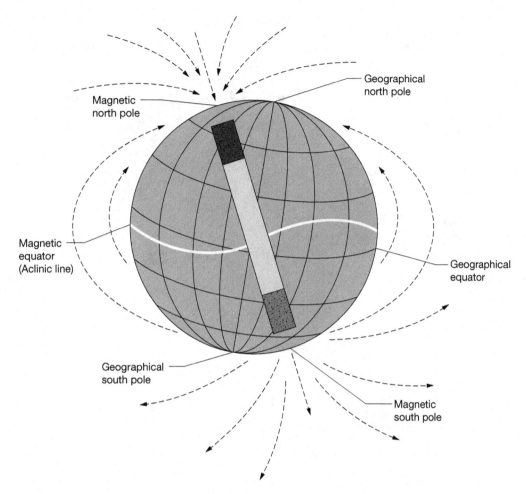

4.6 Earth's magnetic field

Magnetic dip

Referring to Figure 4.7, at the equator, lines of the earth's magnetic field are parallel to the earth's surface; as they approach the magnetic poles they progressively align with the 'bar magnet' and dip downward, until at the poles they will be vertical with respect to the earth's surface. The effect on the bar magnet within a direct-reading magnetic compass is dip, making it unusable in higher latitudes. To compensate for this, the pivot point of the bar magnet is offset from the magnet's centre of gravity (CG, or C of G).

Direct-reading compasses

A simple direct-reading magnetic compass is a mandatory item of equipment on all aircraft, from single-engine trainers through to large transport aircraft – see Figure 4.8. They are constructed with a bar magnet, and a display that has the points of the compass marked on it (Figure 4.9). The bar magnet/display assembly is freely suspended on a pivot and contained within a housing. The compass housing is filled with fluid for damping purposes; this decreases oscillations during turbulence, and decreases friction at the compass pivot point. The fluid typically used is an alcoholic solution, giving rise to the term 'whisky compass'. The bar magnet in the compass is offset from the compass card pivot point to account for dip;

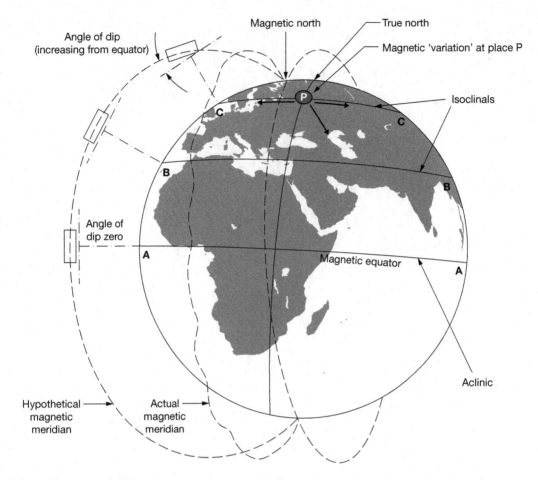

4.7 Earth's magnetic field dip

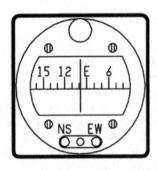

4.8 Direct-reading/standby compass

the centre of gravity of the magnet is behind the pivot point in the northern hemisphere and ahead of it in the southern hemisphere.

Direct-reading magnetic compasses are subject to errors arising when the aircraft is turning and from aircraft acceleration/decelerations. Turning errors (Figure 4.10) are at a maximum when turning onto north or south headings; errors are minimal on east or west headings. Flying on east/west headings produces the largest acceleration/deceleration errors (Figure 4.11) because the bar magnet is always aligned with magnetic north/south; the moment arm between the pivot point and the CG is at its maximum.

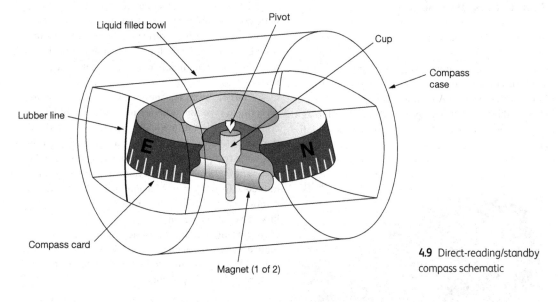

Liquid filled bowl

Pivot

Cup

Compass case

Lubber line

E

N

Compass card

Magnet (1 of 2)

4.9 Direct-reading/standby compass schematic

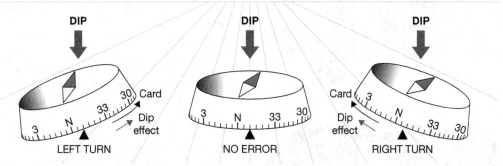

DIP

DIP

DIP

Card

Dip effect

Card

Dip effect

30

33

3

N

3

N

33

30

3

N

33

30

LEFT TURN

NO ERROR

RIGHT TURN

4.10 Turning errors

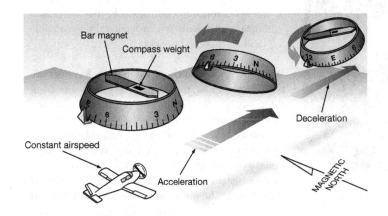

Bar magnet

Compass weight

Deceleration

Constant airspeed

Acceleration

MAGNETIC NORTH

4.11 Acceleration/ deceleration errors

Remote compass system

As previously discussed, magnetic compasses are unreliable in the short term, i.e. during turning manoeuvres. Directional gyroscopes are reliable for azimuth guidance in the short term, but drift over longer time periods. A combined magnetic compass stabilized by a directional gyroscope (referred to as a gyromagnetic compass) can overcome these deficiencies.

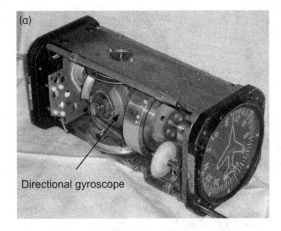

Directional gyroscope

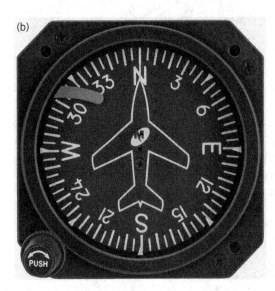

4.12 Directional gyro (a) Internal view (b) DG display

Smaller GA aircraft normally have a single magnetic compass above the instrument panel; this is used in conjunction with the directional gyro instrument – see Figure 4.12. The pilot has to monitor both displays, and then manually correct the DG after each turn.

In larger aircraft, the magnetic compass and directional gyro functions are integrated such that the compass card is automatically driven (or 'slaved') to the magnetic reference. In this arrangement, a remote magnetic compass sensor is installed either in the wing tip or another part of the aircraft where it has minimal influence from sources of magnetic interference. The remote magnetic compass sensor is called a flux gate, flux valve or magnetometer.

Referring to the simplified block diagram in Figure 4.13, the primary components of a remote gyrocompass system are based on the following requirements/components:

1. A magnetic reference; the sensing or detector unit.
2. A gyroscopic reference; normally a directional gyro.
3. An amplifier that provides power to precess the gyro whenever a monitoring signal exists; this is used to synchronize the gyroscopic and magnetic references.
4. A synchronizing indicator; there has to be a means of comparing the gyroscopic and magnetic references, using the difference between them to provide a monitoring signal. This synchronizing indicator provides an indication of the agreement between the gyroscopic and magnetic references.
5. A synchronizing control knob; this provides a means of manual synchronizing the gyroscopic and magnetic references.

When the gyroscopic and magnetic references are both in agreement, there will be no error signal from the amplifier. No monitoring signal is transmitted from the amplifier to precess the gyro unit; the synchronizing indicator shows an agreement between the gyroscopic and magnetic references. When the directional gyro wanders, there is a difference compared to the magnet compass reference and an error signal is produced. The monitoring signal is amplified and used to precess the gyro back to the magnetic reference signal. The gyro is now slowly corrected via precession; during this time, the magnitude of error

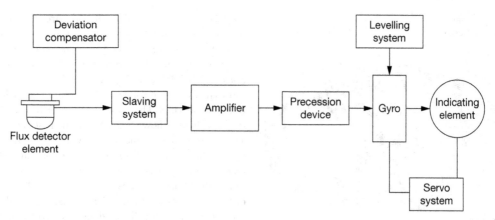

4.13 Remote indicating compass – architecture

signal is indicated via the synchronizing indicator. When the gyroscopic and magnetic references are both back in agreement, there will be no error signal from the amplifier and precession will cease.

> **KEY POINT**
>
> The gyro compass synchronizing indicator shows when the gyroscopic and magnetic references are not in agreement.

When the gyro compass is first powered up, any differences between the gyroscopic and magnetic references will be corrected, as previously described. This can take some time (typically two to three degrees per minute) and so for convenience, provision is made for manual synchronization using the synchronizing knob. This precesses the gyro at a faster rate and is switched off during normal operation.

The directional gyro has been described earlier in the gyroscopic instruments chapter. The magnetic reference sensor or detector unit is based on a magnetometer.

Magnetometer/flux valve

The term 'magnetometer' has widespread meanings; in general it is a measuring instrument used to detect the strength and direction of magnetic fields. The principles of magnetometers are used in geophysical surveys, by the military to detect submarines, in metal detectors and in consumer devices such as mobile phones. The traditional aircraft electromechanical flux valve (or gate) is shown in Figure 4.14.

The flux valve (or flux gate) consists of two main components: the magnetic sensor and the compensating device. The sensor is a pendulous device, free to move in response to pitch and roll, but fixed in yaw. The bowl is partially filled with light oil to dampen oscillations during flight. The compensating device located at the top of the unit consists of two sets of electromagnetic coils. The coils are supplied with a DC voltage from a compass-system amplifier; this is used to compensate the aircraft's magnetic fields.

The flux valve's sensing element is in the form of a three-spoke wheel (Figure 4.15), split through the rim as illustrated to produce horns. The wheel is made from a high-permeability metal, giving a low reluctance path for the earth's magnetic flux. The horns concentrate the earth's magnetic flux. A coil is wound

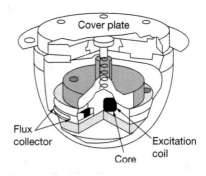

4.14 Electromechanical flux gate

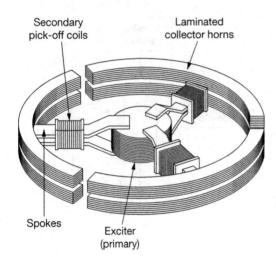

4.15 Flux-gate sensing elements

on each spoke; these 'pick-off coils' are connected together in a star configuration, as shown in Figure 4.16. A separate 'exciter' coil is wound on the centre of the wheel's hub; this is supplied with an AC supply, normally 26 VAC, as in Figure 4.17. This supply will cause an alternating voltage to be produced in each of the pick-off coils. The concentration of the earth's magnetic field is shown flying North and East in Figure 4.18. The earth's magnetic flux will either assist or oppose the effects of the exciter coil; the output of the pick-off coils is directly proportional to the orientation of the spokes/horns with respect to the direction of the earth's magnetic flux. The signal produced by each of the pick-off coils is fed into the gyromagnetic compass system as the magnetic reference.

The physical location of a flux valve (or flux gate) has to be carefully selected; typical locations include

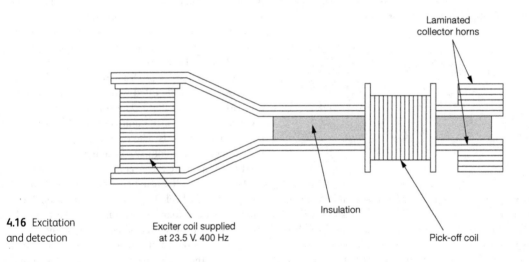

4.16 Excitation and detection

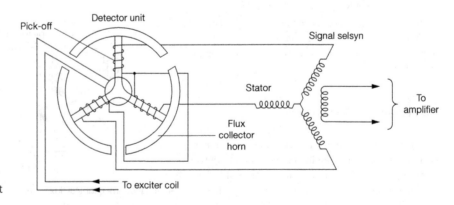

4.17 Flux detector circuit arrangement

(a)

(b)

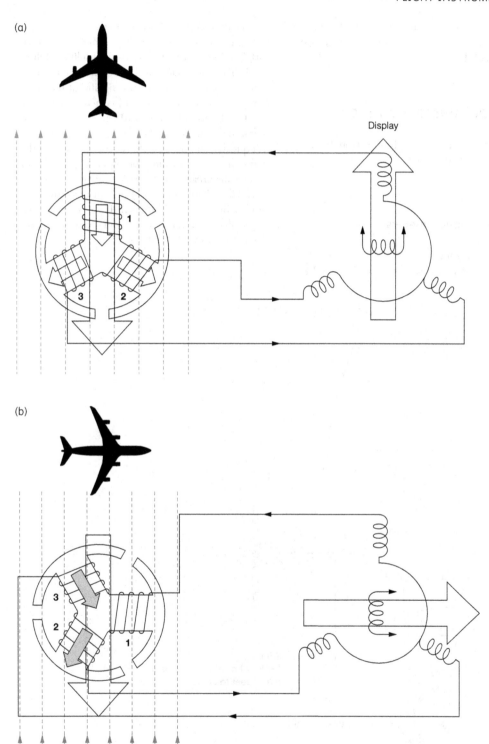

4.18 Flux gate and the earth's magnetic field: (a) Flying north (b) Flying east

the wing tip(s) on aeroplanes and in the boom of a rotorcraft. Further considerations are given in the next section of this chapter.

Solid-state magnetometers

The solid-state magnetometer is an essential part of integrated electronic systems used in AHRS and PFD.

The Garmin GMU 44 is an extremely sensitive three-axis magnetic sensor. It is more sensitive to nearby magnetic disturbances than a traditional flux-gate device. Large electrical loads that are close to the magnetometer can generate significant magnetic interference. For this reason, when selecting a mounting location for the GMU 44, recommended minimum distances must be observed – see Table 4.1 for typical examples. In the event that all of the minimum distances cannot be observed, sources of magnetic disturbances to be avoided are given in order of priority. The chosen location must be surveyed prior to installation of the GMU 44 to verify its acceptability. Typical locations for magnetometers are shown in Figures 4.19 and 4.20.

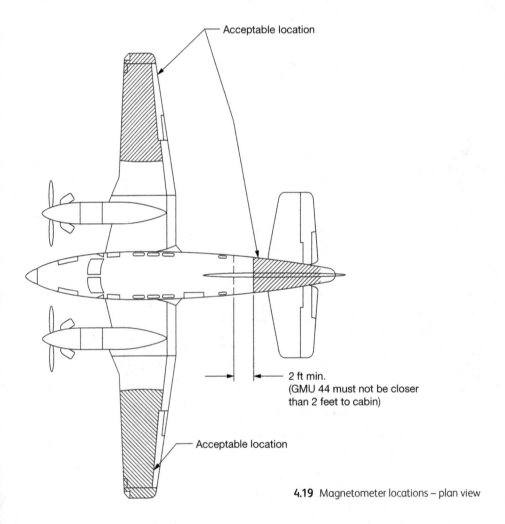

Acceptable location

2 ft min.
(GMU 44 must not be closer than 2 feet to cabin)

Acceptable location

4.19 Magnetometer locations – plan view

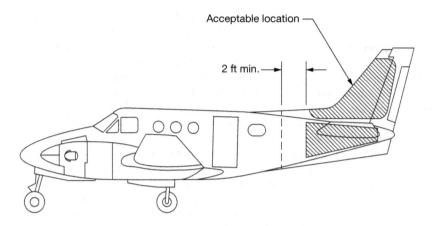

Acceptable location —

2 ft min. —

4.20 Magnetometer locations – side view

Table 4.1 Required distances from magnetic disturbances

Disturbance Source	Priority	Distance*
Electric motors and relays, including servo motors	1	10 ft/3m
Ferromagnetic structure greater than 1kg total (e.g. landing-gear structure)	2	8.2 ft/ 2.5m
Ferromagnetic materials less than 1kg total (e.g. control cables)	3	3 ft/1m
Electrical devices drawing more than 100 mA current	4	3 ft/1m
Electrical conductors passing more than 100 mA current	5	3 ft/1m
Electrical devices drawing less than 100 mA current	6	2 ft/0.6m
Magnetic measuring device (e.g. installed flux gates)	7	2 ft/0.6m
Electrical conductors passing less than 100 mA current	8	3 ft/0.4m

* Garmin recommended minimum distance in feet/metres

In general, wing mounting of the GMU 44 magnetometer is strongly preferred. If wing mounting is not possible, it may be necessary to install the GMU 44 in the tail section of the aircraft. Fuselage mounting is permitted, but not within two feet of the cabin area because of numerous potential disturbances that can interfere with its operation. Ferromagnetic materials can become magnetized and cause magnetic interference. It is important to use nonmagnetic materials to mount the unit. This unit receives power directly from the GRS 77 AHRS (see the next section of this chapter) and communicates with the GRS 77 using an RS-485 digital interface.

If electrical loads are grounded through the airframe, the returning electrical current will flow through the airframe toward the alternator, generator or battery. If the magnetometer lies along this current return path, the current can cause significant magnetic interference. Electrical current return paths are the most common cause of magnetic interference issues. Common examples of this problem are the navigation lights at the wingtips for wing-mounted magnetometers or the strobe light on the tail for vertical stabilizer-mounted magnetometers.

KEY POINT

Non-magnetic materials, e.g. brass screws, are used to mount magnetometers.

KEY POINT

Ferrous-based fasteners are not normally used within 20 inches of magnetometers.

Compass checks

The aircraft's compass system(s) must be checked and/or adjusted for a variety of reasons. Always consult with the aircraft maintenance manual/schedule for a given aircraft type. A basic compass check swing consists of positioning the aircraft on four cardinal headings, each 90 degrees apart, and comparing the actual deviations with the calibrated compass readings. If there are differences outside approved tolerances then a full compass swing is required. Typical reasons for carrying out a check swing are:

1. Whenever the compass accuracy is suspect
2. Following any modification, repair or component replacement involving magnetic materials
3. Following any modification that introduces significant changes to avionic equipment
4. When the freight/cargo contains magnetic materials
5. If the aircraft has been subjected to physical shock, e.g. turbulence or a heavy landing
6. If the aircraft has been subjected to an electrical storm
7. If the aircraft has been in long-term storage (typically for over a year)
8. If the aircraft's operational area is changed to a different geographic location that has a change in magnetic deviation.

The full compass swing requires a more detailed check and calibration procedure; results of the compass swing are kept with the aircraft records and correction values are stated on a placard adjacent to the remote compass – Figure 4.21. In the first instance, a suitable location must be established for the calibration. Many airports and maintenance bases have a dedicated area for compass swings, referred to as a 'compass rose'; this has to be in an area where there is no magnetic interference within a specified area, e.g. power cables, radar transmitters, steel structures, underground pipes and so on. A series of radial lines are marked out, these marks are referenced to magnetic north, and are made at 30- or 45-degree intervals.

The aircraft is positioned on the compass rose and aligned to the radial for magnetic north, with a person in the cockpit running the engine(s) and another person approximately 30 feet in front of the aircraft (facing south) using a calibrated sighting compass. Using hand signals, and/or radio, the person outside the aircraft signals the person in the cockpit to make adjustments to align the aircraft with the compass. The compass's N-S compensator screw is adjusted until the aircraft compass reads North (0°). This process is repeated by moving the aircraft into an E-W direction, and then through a full 360 degrees. As adjustments are made for N-S and E-W, the calibration process takes a difference between readings, leading to a mid-point reading. When the compass swing is completed, the results are filed with the aircraft records, and the calibration card is completed and placed in the approved pilot's full view.

NAVIGATION INDICATORS

Following on from the gyrocompass, navigation displays have evolved into enhanced navigation indicators – the course-deviation indicator (CDI) and horizontal-situation indicator (HSI). A typical CDI is shown in Figure 4.22.

The horizontal situation indicator (HSI) – Figure 4.23 – is located below the artificial horizon/attitude indicator; it is based on the CDI, but with additional features and functions. The HSI combines heading (usually slaved to a remote compass) and VOR/ILS displays, reducing pilot workload. The HSI pointer is set to VOR/ILS course, giving left/right deviations. The HSI is normally interfaced with an autopilot, as described in the autopilot chapter. With the HSI selected to a VOR course, left and right and TO/FROM is indicated by a triangular symbol pointing to the VOR station. With the symbol pointing to the same side as the course-selector arrow, it indicates TO the VOR station, and if it points behind to the side

FOR	N	30°	60°	E	120°	150°
STEER	001°	029°	060°	089°	120°	152°
FOR	S°	210°	240°	W°	300°	330°
STEER	181°	212°	240°	268°	301°	330°

4.21 Compass calibration card

4.22 Course-deviation indicator

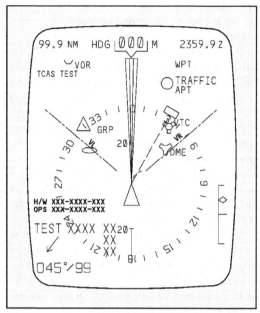

4.24 Electronic horizontal situation indicator (EHSI)

ATTITUDE AND HEADING REFERENCE SYSTEMS

AHRS overview

An attitude and heading reference system (AHRS) consists of sensors on three axes that provide heading, attitude and yaw data for the aircraft. They are designed to replace traditional mechanical gyroscopic flight instruments and provide increased reliability and accuracy.

AHRS consist of laser or MEMS gyroscopes, accelerometers and magnetometers on all three axes. The AHRS will also incorporate integrated processing that calculates attitude and heading solutions. One essential feature of advanced navigation systems is the use of Kalman filters, named after Dr Richard Kalman, who introduced this concept in the 1960s. Kalman filters are optimal recursive data processing algorithms that filter navigation sensor measurements. The mathematical model is based on equations solved by the navigation processor. To illustrate the principles of Kalman filters, consider a navigation system based on inertial navigation sensors with periodic updates from radio-navigation aids (see Chapter 5). One key operational aspect of inertial navigation is that system

4.23 Horizontal situation indicator

opposite the course selector, it indicates FROM the station. HSIs for a larger aircraft will include more display functions and interfaces. An electronic HSI (EHSI) is illustrated in Figure 4.24.

errors accumulate with time. When the system receives a position fix from navigation aids, the inertial navigation system's errors can be corrected.

The key feature of the Kalman filter is that it can analyse these errors and determine how they might have occurred; the filters are recursive, i.e. they repeat the correction process on a succession of navigation calculations and can 'learn' about the specific error characteristics of the sensors used. The numerous types of navigation sensors employed in RNAV systems vary in their principle of operation, as described in the relevant chapters of this book. Kalman filters take advantage of the dissimilar nature of each sensor type; with repeated processing of errors, complementary filtering of sensors can be achieved.

AHRS differs from traditional inertial-navigation systems by attempting to estimate only attitude (i.e. roll, pitch and yaw) states, rather than heading, position and velocity, as is the case with an INS. AHRS are in widespread use in commercial and business aircraft and are typically integrated with electronic flight-information systems (EFIS) to form the primary flight display. AHRS can also be integrated with air-data computers to form 'air data, attitude and heading reference systems' (ADAHRS).

Attitude and heading reference system

The attitude and heading reference system (AHRS) combine inputs from navigation sensors, e.g. GPS, magnetometer and an air-data computer. The AHRS provides an accurate digital output and referencing of aircraft position, rate, vector and acceleration data. It is able to restart and properly realign itself while the aircraft is moving. Garmin's GRS 77 AHRS has the following physical specifications:

- Height: 3.25 inches (8.36 cm)
- Width: 3.75 inches (9.53 cm)
- Length: 8.5 inches (21.59 cm)
- Weight: 2.40 lb (1.08 kg)
- Voltage Range: 10–33 VDC

GRS 77 AHRS Performance:

- Bank/Pitch Error: $\pm 1.25°$ within 30° roll, left or right and 15° pitch, nose up or down
- Manoeuvres Range: 360° pitch and roll

- Rotation Rate: $\pm 200°$ per second
- Heading: $\pm 2°$ straight and level flight

ELECTRONIC FLIGHT DISPLAYS

Introduction

Electronic flight displays can either be installed as retrofit, e.g. the Aspen EFD range, or installed during production of the aircraft, e.g. the Garmin 1000 system.

Garmin's G1000 is an all-glass avionics suite designed for original equipment manufacturer (OEM) installation on a wide range of business aircraft. It is an integrated package that makes flight information easier to scan and process. G1000's design brings new levels of situational awareness, simplicity and safety to the cockpit. The G1000 synthetic vision technology is described in more detail in a subsequent chapter.

Aspen Avionics' Evolution Flight Display (EFD) System

This is a flexible, expandable and upgradable electronic flight-instrument system (EFIS) available for general aviation aircraft, designed to replace traditional mechanical primary flight instruments, in whole or in part, all at once or in phases. This modularity and upgradability allows the system to grow with the pilot and the airplane, over time and affordably. The EFD system is built around the EFD1000 PFD, which replaces a vertical pair of the six primary flight instruments. The PFD has a bright, high-resolution, six-inch diagonal liquid-crystal display (LCD) and a number of knobs, buttons and keys the pilot uses to control the system. The three-inch diameter, four-inch deep can on the back of the display slides into existing panel cut-outs (where the top mechanical instrument used to be) – see Figure 4.25.

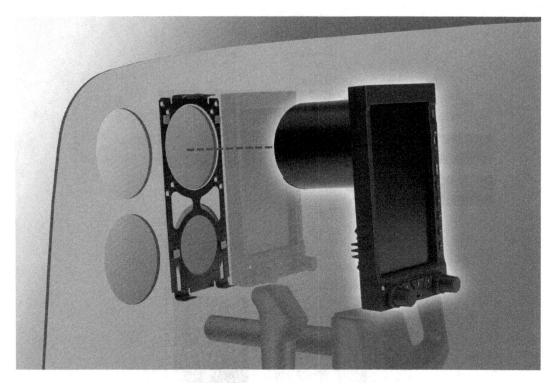

4.25 EFD1000 retrofit principles

The centre of the EFD System is the EFD1000 Primary Flight Display (PFD), which replaces the traditional mechanical attitude indicator (AI) and directional gyro (DG) or horizontal situation indicator (HSI) – Figure 4.26. The PFD system typically consists of four components:

1. EFD1000 primary flying display unit (PFD)
2. Configuration module (CM)
3. Remote-sensor module (RSM)
4. Analogue converter unit (ACU)

The ACU converts older analogue signals and interfaces to the industry-standard digital ARINC 429 interface, which is the standard means of communication of the PFD. In some installations, generally when the aircraft is not equipped with an autopilot and has only digital GPS/Nav/Comm, the ACU may be omitted.

The system architecture in Figure 4.27 shows the relationships of the PFD, RSM, CM and ACU. The primary flight-display (PFD) unit is a digital system that consists of a high resolution, six-inch diagonal

4.26 EFD1000 graphics

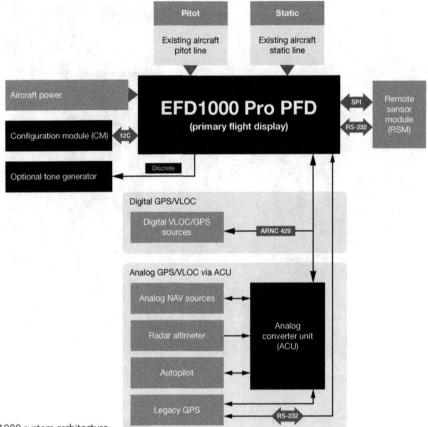

4.27 EFD1000 system architecture

colour LCD display, user controls, photocell and microSD data-card slot. The three-inch diameter, four-inch deep can on the back of the display contains a non-removable electronics module that includes:

- A sensor board with solid-state attitude and heading reference system (AHRS) and digital air-data computer (ADC)
- A main application processor (MAP) board with central processing unit (CPU), graphics processor and system memory
- An input-output processor (IOP) board for integrating communications with other aircraft systems.

Also on the rear of the unit are:

- An access cover for removing and replacing the built-in backup battery
- Pneumatic connections to the aircraft's pitot and static systems

- A 44-pin D-sub connector for electrical connections to the PFD
- A cooling fan, to cool the electronics and LCD backlights.

The PFD mounts to the front surface of the instrument panel.

The configuration module (CM) contains an erasable electronic programmable read-only memory (EEPROM) device that retains system configuration and calibration data and provides two primary functions:

- Retains aircraft-specific configuration information, calibration data and user settings, allowing the PFD to be swapped for service purposes without re-entering or recalibrating the installation
- Contains a licence key that configures the PFD software features.

The CM is typically attached to the wire bundle coming out of the D-sub connector on the display unit. The analogue converter unit (ACU) included with the PFD system enables the all-digital EFD1000 System to interface with analogue avionics when required. The ACU converts multiple analogue interfaces to the digital ARINC 429 buses supported by the PFD. Control parameters, such as desired heading, are also sent from the PFD to the ACU for conversion to analogue format for autopilot support. The ACU is required when any of the following capabilities are required in a PFD installation:

- Interface to supported autopilots
- Interface to conventional VHF navigation radios
- Interface to legacy (non-ARINC 429) GPS navigators
- Interface to supported radar altimeter decision height annunciations

If ARINC 429-based digital radios, such as the Garmin 400/500-series GPS, navigation, and communications radios, are installed in the aircraft, and no other aircraft interfaces are desired, the ACU is not required.

The remote-sensor module (RSM) is an integral part of the PFD system and works together with the display-unit sensors as part of the AHRS and ADC. The RSM looks and mounts like a GPS antenna and is mounted on the exterior of the fuselage, typically aft of the cabin. The RSM contains the following sub-systems:

- 3D magnetic flux (heading) sensors
- Outside air temperature (OAT) sensor
- Emergency backup GPS engine and antenna

The RSM communicates with the PFD via a digital cable connection. The PFD is a flat-panel LCD primary flight instrument that presents the pilot with all of the information from the traditional six-pack of mechanical instruments: airspeed, attitude, altitude, turn coordinator, heading indicator (or HSI) and vertical-speed indicator (VSI). Modern technology and standard EFIS symbology enable the consolidation of all six instruments into a single display, tightening the pilot's instrument scan and reducing pilot workload.

The PFD is a single vertical instrument that replaces the existing attitude indicator and heading indicator/HSI. The display is divided into three parts: an upper attitude display, a lower navigation display and a data bar between the upper and lower halves. The attitude and navigation displays are highly

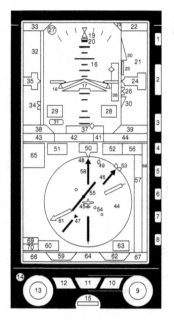

4.28 EFD display area

customizable – from stripped-down, minimalist presentations, to dense, information-rich displays – depending on pilot preference and phase of flight. This section gives an overview of all the instruments, information and controls of the Aspen PFD – see Tables 4.2 through 4.5, and Figure 4.28.

Garmin 1000 system

This section gives an overview of the Garmin G1000 integrated cockpit system as installed in a range of GA aircraft; Figure 4.29 shows the King Air installation. The G1000 offers a wealth of flight-critical data at the pilot's fingertips. Its glass flight deck presents flight instrumentation, navigation, weather, terrain, traffic

4.29 G1000 King Air

Table 4.2 Aspen EFD controls

Controls	
1. Reversion and power button	
2. Range buttons	
3. Menu button	
4. 1/2 Hot key menu 1 of 2	2/2 Hot key menu 2 of 2
5. MIN – Minimums on/off	No function
6. 360/ARC View	No function
7. GPS steering on/off	TRFC – traffic overlay on/off
8. BARO	BARO
9. Right knob	
10. Lower right button, double-line bearing pointer source select	
11. CDI navigation source select button	
12. Lower left button, single-line bearing pointer source select	
13. Left knob	
14. Automatic dimming photocell	
15. microSD card slot	

Table 4.3 Aspen EFD attitude display

Attitude display
16. Attitude display
17. Aircraft reference symbol
18. Single-cue flight director 2
19. Roll pointer
20. Slip/skid indicator
21. Altitude tape
22. Selected altitude field (controls the altitude bug)
23. Altitude alerter
24. Numerical altitude indication, altitude drum/pointer
25. Altitude trend vector
26. Altitude bug
27. Decision height annunciation 3, 4
28. Selected minimums field
29. Radio altitude 3, 4
30. Minimums marker
31. Lateral deviation indicator (LDI) navigation source indication
32. Airspeed tape
33. Selected airspeed field (controls the airspeed bug)
34. Airspeed bug
35. Numerical airspeed indicator, airspeed drum/pointer
36. Vertical deviation indicator (VDI)
37. Lateral deviation indicator (LDI)

Table 4.4 Data bar

Data bar
38. True airspeed (TAS) or Mach number
39. Barometric pressure setting field
40. Wind direction and speed
41. Wind direction arrow
42. Outside air temperature (OAT)
43. Ground speed

Table 4.5 Navigation display

Navigation display
44. Navigation display
45. Own-ship symbol
46. Course pointer
47. TO/FROM indicator
48. Rate of turn indicator
49. Ground track indicator
50. Numerical direction indicator
51. Selected course (CRS) field
52. Selected heading field
53. Heading bug
54. Course deviation scale
55. Course deviation indicator
56. Vertical speed numerical value
57. Vertical speed tape
58. Single-line bearing pointer
59. Single-line bearing pointer source
60. Single-line bearing pointer source info block
61. Double-line bearing pointer
62. Double-line bearing pointer source
63. Double-line bearing pointer source info block
64. Selected CDI navigation source
65. Selected CDI navigation source information block
66. Left knob state
67. Right knob state
68. Hot key labels
69. Base-map range
70. De-clutter level

and engine data on large-format, high-resolution displays.

The G1000 adapts to a broad range of aircraft models. It can be configured as a 2-display or 3-display system, with a choice of 10" or 12" flat-panel LCDs interchangeable for use as either a primary flight display (PFD) or multi-function display (MFD). An optional 15" screen is also available for even larger format MFD configurations.

The G1000 replaces traditional mechanical gyroscopic flight instruments with Garmin's GRS77 attitude and heading reference system (AHRS). AHRS provides accurate, digital output and referencing of the aircraft's position, rate, vector and acceleration data. It is able to restart and properly reference itself while the aircraft is moving.

G1000 also includes the GFC 700 autopilot. The GFC 700 is capable of using all data available to G1000 to navigate, including the ability to maintain airspeed references and optimize performance over the entire airspeed envelope.

G1000 integrates built-in terrain and navigation databases, providing a clear, concise picture for situational awareness. A database supports onscreen navigation, communication and mapping functions, including an overlay of the aircraft's position on the electronic approach chart to provide a visual cross-check. Using information from the built-in terrain and obstacles databases, G1000 displays colour coding to graphically alert the pilot when proximity conflicts loom ahead. In addition, the pilot can augment G1000 with optional Class-B terrain awareness and warning system (TAWS) for an extra margin of safety in the air.

The pilot can also view position on taxiways with a built-in database of airport diagrams. Optional features let the pilot quickly find and view departure procedures (DP), standard terminal arrival routes (STARs), approach charts and airport diagrams on the MFD.

For added visual orientation, optional Garmin synthetic vision technology (SVT™) is now available. Using sophisticated graphics modelling, the system recreates a 3D 'virtual reality' database landscape on the pilot and co-pilot PFDs. Thus, Garmin SVT enables the pilot to clearly visualize nearby flight and en-route navigation features – even in solid IFR or night-time VFR conditions.

With an optional subscription to XM WX Satellite Weather™ and the addition of the GDL 69 or 69A data link receiver, the pilot will have access to high-resolution weather for the US, right in the cockpit. The GDL 69 provides NEXRAD, METARs, TAFs, lightning and more that can be laid directly over topographic map databases. For global weather information, an optional GSR 56 can connect the pilot to the Iridium satellite network, which also provides voice and text messaging connectivity, as well as position reporting. For the best in 'scan-your-own' weather analysis, the G1000 can be interfaced with Garmin's GWX 68™ digital colour radar. By adding a Garmin Mode S transponder, the G1000 will also

display traffic-information services (TIS) alerts that identify surrounding air traffic. The G1000 system includes the following line-replaceable units (LRUs):

- GDU 1040 Primary Flight Display (PFD)
- GDU 1040 Multi-Function Display (MFD)
- GIA 63 Integrated Avionics Units (2)
- GEA 71 Engine/Airframe Unit
- GDC 74A Air-Data Computer (ADC)
- GRS 77 Attitude and Heading Reference System (AHRS)
- GMU 44 Magnetometer

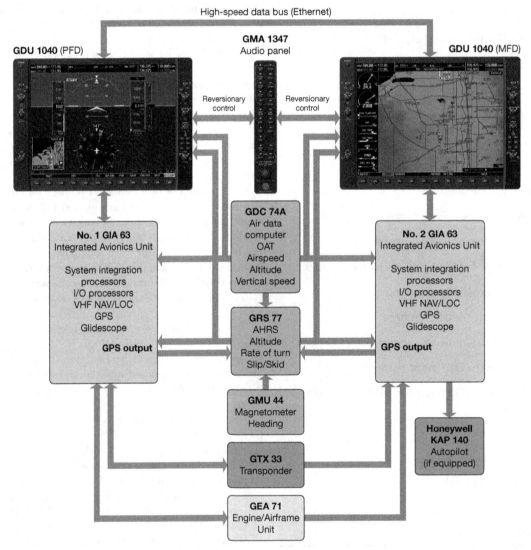

4.30 G1000 top-level interfaces

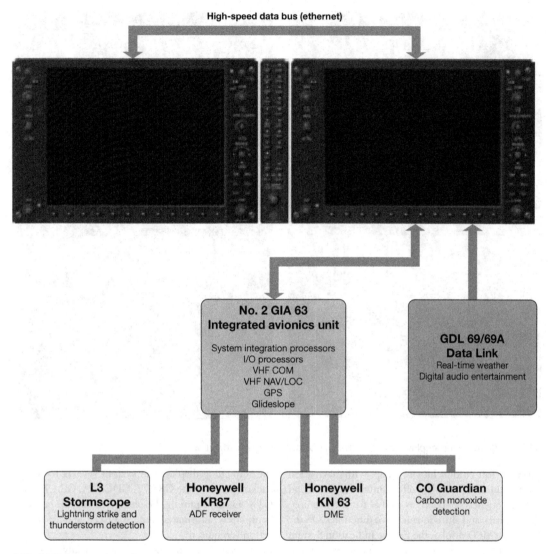

4.31 G1000 optional interfaces

- GMA 1347 Audio System with integrated Marker Beacon Receiver
- GTX 33 Mode-S Transponder
- GDL 69/69A Data Link

The LRUs are further described in the following section. All LRUs have a modular design, which eases troubleshooting and maintenance of the G1000 system. A top-level G1000 block diagram is given in Figure 4.30. Additional or optional interfaces are depicted in Figure 4.31.

Normal mode

The PFD and MFD are connected together on a single Ethernet bus, allowing for high-speed communication between the two units. Each GIA 63 is connected to a single display, which allows the units to share information, thus enabling true system integration. In normal operating mode, the PFD displays graphical flight instrumentation in lieu of the traditional gyro instruments. Attitude, heading, airspeed, altitude and vertical speed are all shown on one display. The MFD

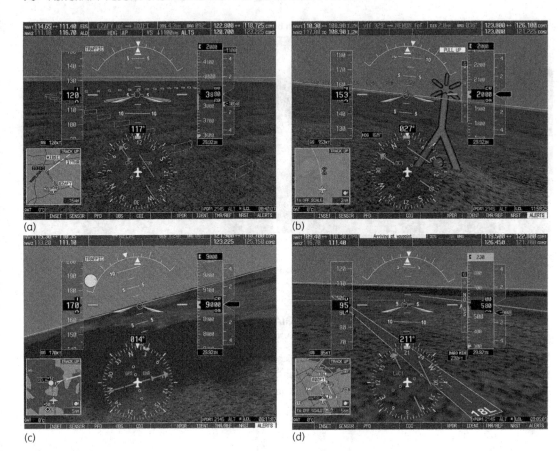

(a)   (b)

(c)   (d)

4.32 G1000 functional displays: (a) Pathways (b) Obstacles (c) Traffic (d) Runway

shows a full-colour moving map with navigation information. Both displays offer control for COM and NAV frequency selection, as well as for the heading, course/baro and altitude-reference functions. On the left of the MFD display, the Engine Indication System (EIS) cluster shows engine and airframe instrumentation. Typical G1000 functional modes are illustrated in Figure 4.32.

Reversionary mode

Should a failure occur in either display, the G1000 automatically enters reversionary mode. In reversionary mode, all important flight information is shown on the remaining display. If a display fails, the GIA 63-GDU 1040 Ethernet interface is cut off. Thus, the GIA can no longer communicate with the remaining display, and the NAV and COM functions provided to the failed display by the GIA are flagged as invalid on

the remaining display. The system reverts to using backup paths for the GRS 77, GDC 74A, GEA 71 and GTX 33, as required. The change to backup paths is completely automated for all LRUs, and no pilot action is required.

Reversionary mode may also be manually activated by the pilot if the system fails to detect a display problem. Reversionary mode is activated manually by pressing the red DISPLAY BACKUP button at the bottom of the Audio Panel. Pressing this button again deactivates reversionary mode.

ELECTRONIC FLIGHT BAGS

An electronic flight bag (EFB) is an electronic display system intended for flight crew (or cabin crew members), replacing information and data traditionally based on paper documents and manuals (e.g. navigation charts, operating manuals and so on). The

EFB may also support other functions that have no paper equivalent, e.g. data communication and other systems. A basic EFB can perform flight-planning calculations and display a variety of navigational charts, operations manuals, aircraft checklists and so on. Advanced EFBs are fully certified and integrated with aircraft systems. EFBs are classified in one of three ways: Class 1, 2 or 3.

Class 1 EFB systems are typically standard commercial off-the-shelf (COTS) equipment such as laptops or handheld electronic devices. These devices are deemed as portable or loose equipment and are typically stowed during critical phases of flight. Class 2 EFB systems can either be commercial off-the-shelf (COTS) equipment or designed for the purpose. They are typically mounted in the aircraft, with the display being viewable to the pilot during all phases of flight. Under certain conditions, they can be certified to interface with aircraft systems. They do not share any display or other input/output device (e.g. keyboard, pointing device) with certified aircraft systems. Class 2 EFBs do not require any installation/removal tools. Class 3 EFBs are neither Class 1 nor 2; they are installed items of equipment with certified operating-system software. A Class 3 EFB is part of the certified aircraft configuration – including installed applications, database, resources and so on.

A typical Class 1/2 EFB is the Garmin 'GPSMAP 696'; this is an all-in-one EFB designed exclusively for aviation. Featuring a 7" diagonal high-definition sunlight-readable display screen, detailed electronic charts and weather data, the 696 is a pilot's personal avionics system. Referring to Figure 4.33, the bezel around the screen includes soft keys and a joystick control. The internal GPS provides position updates five times per second, allowing the 696 to present flight data smoothly and continuously.

With optional modes, the GPSMAP 696 provides key functions of a Class 1/Class 2 electronic flight bag (EFB), reducing the use of paper charts in the cockpit. The 696 can display vector airways, jet routes, minimum en-route altitudes and leg distance, standard terminal arrival routes (STARs), approach charts and airport diagrams replicating traditional en-route charts.

MULTIPLE-CHOICE QUESTIONS

1. For a direct-reading compass, the angle of dip is greatest at:

    (a) The equator
    (b) East/west headings
    (c) The magnetic poles

2. This basic 'six pack' configuration is arranged with:

    (a) Top row: airspeed, artificial horizon, altimeter; bottom row: radio compass, direction indicator, vertical speed
    (b) Top row: vertical speed, artificial horizon, altimeter; bottom row: airspeed, radio compass, direction indicator
    (c) Top row: direction indicator, vertical speed, artificial horizon; bottom row: airspeed, radio compass, altimeter

3. Acceleration errors in direct-reading compasses are

    (a) Minimum on east/west headings
    (b) Maximum on north/south headings
    (c) Maximum on east/west headings

4.33 Electronic flight bag

4. In reversionary mode, all important PFD flight information is shown on the:

   (a) Electronic Flight Bag (EFB)
   (b) Remaining display
   (c) AHRS

5. The gyrocompass synchronizing indicator shows when the gyroscopic and magnetic references are:

   (a) Not in agreement
   (b) In agreement
   (c) Powered off

6. The PFD is a single vertical instrument that replaces the existing

   (a) Altimeter/VSI/ASI
   (b) Electronic Flight Bag (EFB)
   (c) Attitude Indicator and Heading Indicator/ HSI

7. Magnetic interference caused by the aircraft's structure and components is called:

   (a) Deviation
   (b) Variation
   (c) Magnetic dip

8. Class 1 EFB systems are typically:

   (a) Standard commercial off-the-shelf devices
   (b) Fully certified and integrated with aircraft systems
   (c) Viewable to the pilot during all phases of flight

9. The flux-gate sensor is a pendulous device, free to move in response to:

   (a) Pitch and yaw, but fixed in roll
   (b) Pitch and roll, but fixed in yaw
   (c) All three axes, pitch, roll and yaw

10. The 'flying T' configuration is arranged with the:

   (a) Artificial horizon/attitude indicator in the top/centre, air-speed indicator (ASI) to the right, altimeter to the left and directional gyro underneath
   (b) Artificial horizon/attitude indicator in the top/centre, air-speed indicator (ASI) to the left, altimeter to the right and directional gyro underneath
   (c) Directional gyro in the top/centre, air-speed indicator (ASI) to the left, altimeter to the right and artificial horizon/attitude indicator underneath

5 Navigation

This chapter gives an overview of aircraft navigation and is complimentary to the other chapters in the book. Navigation is a process of determining a current position and direction of travel; for aircraft, this occurs in three dimensions. With knowledge of the current location and ultimate destination, the journey's progress has to be checked along the way. Finding a position on the earth's surface and deciding on the direction of travel can be simply made by observations or by mathematical calculations. Aircraft navigation is no different, except that the speed of travel is much faster! Navigation systems for aircraft have evolved with the nature and role of the aircraft itself. Starting with visual references and the basic compass, leading onto radio ground aids and self-contained systems, many techniques and methods are employed.

Although the basic requirement of a navigation system is to guide the crew from point A to point B, increased traffic density and airline economics means that more than one aircraft is planning a specific route. Flight planning takes into account such things as favourable winds, popular destinations and schedules. Aircraft navigation is therefore also concerned with the management of traffic and the safe separation of aircraft. This chapter reviews some basic features of the earth's geometry as it relates to navigation, and introduces some basic aircraft navigation terminology.

TERRESTRIAL NAVIGATION

Before looking at the technical aspects of navigation systems, we need to review some basic features of the earth and examine how these features are employed for aircraft navigation purposes. Although we might consider the earth to be a perfect sphere, this is not the case. There is a flattening at both the poles, such that the earth is shaped more like an orange. For short distances, this is not significant; however, for long-range (i.e. global) navigation we need to know some accurate facts about the earth. The mathematical definition of a sphere is where the distance (radius) from the centre to the surface is equidistant. This is not the case for the earth, where the actual shape is referred to as an oblate spheroid.

> **KEY POINT**
>
> Although we might consider the earth to be a perfect sphere, this is not the case. The actual shape of the earth is referred to as an oblate spheroid.

> **KEY POINT**
>
> Longitude referenced to the prime meridian extends east or west up to 180 degrees. Latitude is the angular distance north or south of the equator; the poles are at latitude of 90 degrees.

KEY POINT

The nautical mile (unlike the statute mile) is directly linked to the geometry of the earth. This quantity is defined by distance represented by one minute of arc of a great circle (assuming the earth to be a perfect sphere).

KEY POINT

Both latitude and longitude are angular quantities measured in degrees. For accurate navigation, degrees can be divided by 60, giving the unit of 'minutes'; these can be further divided by 60, giving the unit of 'seconds'.

Position

To define a unique two-dimensional position on the earth's surface, a coordinate system using imaginary lines of latitude and longitude is drawn over the globe, as seen in Figure 5.1. Lines of longitude join the poles in great circles or meridians. A great circle is defined as the intersection of a sphere by a plane passing through the centre of the sphere; this has a radius

measured from the centre to the surface of the earth. These north–south lines are spaced around the globe and measured in angular distance from the zero (or prime) meridian, located in Greenwich, London. Longitude referenced to the prime meridian extends east or west up to 180 degrees. Note that the distance between lines of longitude converge at the poles. Latitude is the angular distance north or south of the equator; the poles are at latitude 90 degrees.

For accurate navigation, the degree (symbol ° after the value, e.g. 90° north) is divided by 60, giving the unit of minutes (using the symbol ' after numbers), e.g. one half of a degree will be 30'. This can be further refined into smaller units by dividing again by 60 giving the unit of seconds (using the symbol " after numbers), e.g. one half of a minute will be 30". A second of latitude (or longitude at the equator) is approximately 31 metres, just over 100 feet. A unique position on the earth's surface, e.g. Land's End in Cornwall, UK, using latitude and longitude is defined as:

Latitude N 50° 04' 13" Longitude W 5° 42' 42"

Direction

Direction to an observed point (bearing) can be referenced to a known point on the earth's surface, e.g. magnetic north. Bearing is defined as the angle between the vertical plane of the reference point

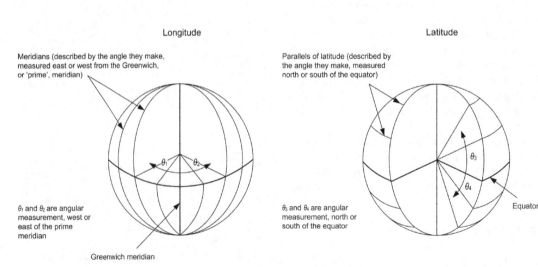

5.1 Longitude and latitude

through to the vertical plane of the observed point. Basic navigational information is expressed in terms of compass points from zero referenced to north through 360° in a clockwise direction – see typical compass displays, Figure 5.2. For practical navigation purposes, north has been taken from the natural feature of the earth's magnetic field; however, magnetic north is not at 90° latitude; the latter defines the position of true north. The location of magnetic north is in the Canadian Arctic, approximately 83° latitude and 115° longitude west of the prime meridian; see the location of magnetic north, Figure 5.3. Magnetic north is a natural feature of the earth's geology; it is slowly drifting across the Canadian Arctic at approximately 40 km northwest per year. Over a long period of time, magnetic north describes an elliptical path. The Geological Survey of Canada keeps track of this motion by periodically carrying out magnetic surveys to redetermine the pole's location. In addition to this long-term change, the earth's magnetic field is also affected on a random basis by the weather, i.e. electrical storms.

5.3 Location of magnetic north

5.2 Compass indications

Navigation charts based on magnetic north have to be periodically updated to consider this gradual drift. Compass-based systems are referenced to magnetic north; since this is not at 90° latitude there is an angular difference between magnetic and true north. This difference will be zero if the aircraft's position happens to be on the same longitude as magnetic north, and maximum at longitudes ± 90° either side of this longitude. The angular difference between magnetic north and true north is called magnetic variation. It is vital that when bearings or headings are used, we are clear on what these are referenced to.

The imaginary lines of latitude and longitude described above are curved when superimposed on the earth's surface; they also appear as straight lines when viewed from above. The shortest distance between points A and B on a given route is a straight line. When this route is examined, the projection of the path (the track) flown by the aircraft over the earth's surface is described by a great circle.

Flying in a straight line implies that we are maintaining a constant heading, but this is not the case. Since the lines of longitude converge, travelling at a constant angle at each meridian yields a track that actually curves as illustrated by flying a constant

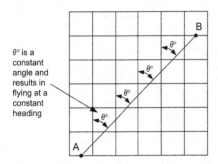

$\theta°$ is a constant angle and results in flying at a constant heading

(a) Local meridians and the rhumb line

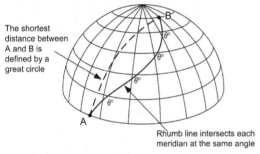

The shortest distance between A and B is defined by a great circle

Rhumb line intersects each meridian at the same angle

(b) Great circle and the rhumb line

5.4 Rhumb line

heading, or rhumb line, (see Figure 5.4). A track that intersects the lines of longitude at a constant angle is referred to as a rhumb line. Flying a rhumb line is readily achieved by reference to a fixed point, e.g. magnetic north. The great circle route, however, requires that the direction flown (with respect to the meridians) changes at any given time, a role more suited to a navigation computer.

Distance and speed

The standard unit of measurement for distance used by most countries around the world (the exceptions being the UK and USA) is the kilometre (km). This quantity is linked directly to the earth's geometry; the distance between the poles and equator is 10,000 km. The equatorial radius of the earth is 6,378 km; the polar radius is 6,359 km.

For aircraft navigation purposes, the quantity of distance used is the nautical mile (nm). This quantity is defined by distance represented by one minute of

arc of a great circle (assuming the earth to be a perfect sphere). The nautical mile (unlike the statute mile) is therefore directly linked to the geometry of the earth. Aircraft speed, i.e. the rate of change of distance with respect to time, is given by the quantity 'knots' – nautical miles per hour.

Calculating the great circle distance between two positions defined by an angle is illustrated in Figure 5.5. The distance between two positions defined by their respective latitudes and longitudes, (lat1, lon1) and (lat2, lon2), can be calculated from the formula:

$$d = cos^{-1}(sin(lat1) \times sin(lat2) + cos(lat1) \times cos(lat2) \times cos(lon1 - lon2))$$

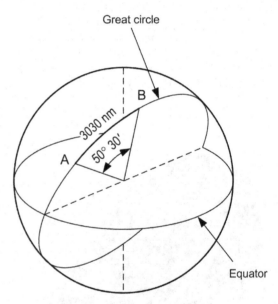

Great circle

Equator

$50° \ 30' = (50 \times 60) + 30 = 3030' = 3030 \ nm$

5.5 Great circle distances

DEAD RECKONING

Estimating a position by extrapolating from a known position and then keeping note of the direction, speed and elapsed time is known as dead reckoning. An aircraft passing over a given point on a heading of 90° at a speed of 300 knots will be five miles due east of the given point after one minute – if, that is, the aircraft is flying in zero wind conditions. In realistic

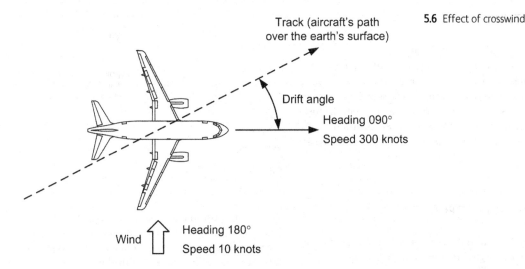

Track (aircraft's path over the earth's surface)

Drift angle

Heading 090°
Speed 300 knots

Heading 180°
Speed 10 knots

Wind

5.6 Effect of crosswind

terms, the aircraft will almost certainly be exposed to wind at some point during the flight and this will affect the navigation calculation. With our aircraft flying on a heading of 90° at a speed of 300 knots, let's assume that the wind is blowing from the south (as a crosswind) at 10 knots, as in Figure 5.6. In a one-hour time period, the air that the aircraft is flying in will have moved north by ten nautical miles. This means that the aircraft's path (referred to as its track) over the earth's surface is not due east. In other words, the aircraft track is not the same as the direction in which the aircraft is heading. This leads to a horizontal displacement (drift) of the aircraft from the track it would have followed in zero wind conditions.

The angular difference between the heading and track is referred to as the drift angle (quoted as being to port/left or starboard/right of the heading). If the wind direction is in the same direction as the aircraft is heading, i.e. a tail wind, the aircraft speed of 300 knots through the air will equate to a ground speed of 310 knots. Likewise, if the wind is from the east (a headwind) the ground speed will be 290 knots.

Knowledge of the wind direction and speed allows the crew to steer the aircraft into the wind such that the wind actually moves the aircraft onto the desired track. For dead-reckoning purposes, we can resolve these figures in mathematical terms and determine an actual position by triangulation, as illustrated in Figure 5.7. Although the calculation is straightforward, the accuracy of navigation by dead reckoning will depend on up-to-date knowledge of wind speed and direction. Furthermore, we need accurate measurements of

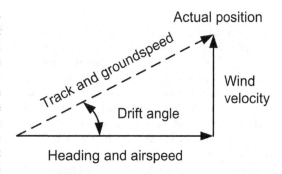

Actual position

Track and groundspeed

Wind velocity

Drift angle

Heading and airspeed

5.7 Resolving actual position

speed and direction. Depending on the accuracy of measuring these parameters, positional error will build up over time when navigating by dead reckoning. We therefore need a means of checking our calculated position on a periodic basis; this is the process of position fixing.

KNOWLEDGE POINT

Dead reckoning is used to estimate a position by extrapolating from a known position and then keeping note of the direction and distance travelled.

POSITION FIXING

When travelling short distances over land, natural terrestrial features such as rivers, valleys, hills and so on can be used as direct observations to keep a check on (pinpointing) the journey's progress. If the journey is by sea, we can use the coastline and specific features such as lighthouses to confirm our position. If the journey is made at night or out of sight of the coast, we need other means of fixing our position.

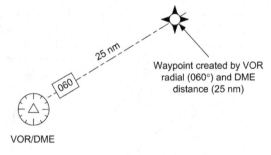

Waypoint created by VOR radial (060°) and DME distance (25 nm)

VOR/DME

5.8 Position fixing

The early navigators used the sun, stars and planets very effectively for navigation purposes; if the position of these celestial objects is known, then the navigator can confirm a position anywhere on the earth's surface. Celestial navigation (or astronavigation) was used very effectively in the early days of long-distance aircraft navigation. Indeed, it has a number of distinct advantages when used by the military: the aircraft does not radiate any signals; navigation is independent of ground equipment; the references cannot be jammed; navigation references are available over the entire globe.

The disadvantage of celestial navigation for aircraft is that the skies are not always clear and it requires a great deal of skill and knowledge to be able to fix a position whilst travelling at high speed. Although automated celestial navigation systems were developed for use by the military, they are expensive; modern avionic equipment has now phased out the use of celestial navigation for commercial aircraft.

The earliest ground-based references (navigation aids) developed for aircraft navigation are based on radio beacons. These beacons can provide angular and/or distance information; when using this information to calculate a position fix, the terms are referred to mathematically as theta (θ) and rho (ρ). By utilising the directional properties of radio waves, the intersection of signals from two or more navigation aids can be used to fix a position (theta–theta), as seen in Figure 5.8. Alternatively, if we know the distance and direction (bearing) to a navigation aid,

the aircraft position can be confirmed (rho–theta). Finally, we can establish our position if we know the aircraft's distance (rho–rho) from any two navigation aids, i.e. without knowledge of the bearing.

MAPS AND CHARTS

Maps provide the navigator with a representative diagram of an area that contains a variety of physical features, e.g. cities, roads and topographical information. Charts contain lines of latitude and longitude, together with essential data such as the location of navigation aids. Creating charts and maps requires that we transfer distances and geographic features from the earth's spherical surface onto a flat piece of paper. This is not possible without some kind of compromise in geographical shape, surface area, distance or direction. Many methods of producing charts have been developed over the centuries; the choice of projection depends on the intended purpose.

In the sixteenth century, the Flemish mathematician, geographer and cartographer Gerhardus Mercator (1512–1594) developed what was to become the standard chart format for nautical navigation. This is a cylindrical map projection where the lines of latitude and longitude are projected from the earth's centre: the Mercator projection (see Figure 5.9). Imagine a cylinder of paper wrapped around the globe and a light inside the globe; this projects the lines of latitude and longitude onto the paper. When the cylinder is unwrapped, the lines of latitude appear incorrectly as having equal length. Directions and the shape of geographic features remain true; however, distances and sizes become distorted. The advantage of using this type of chart is that the navigator sets a constant heading to reach the destination. The

Central meridian selected by cartographer

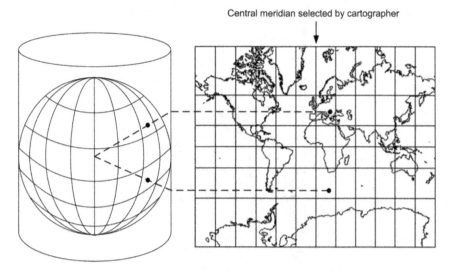

5.9 Mercator projection

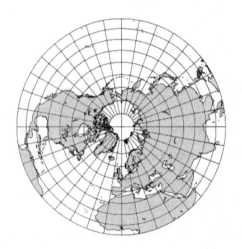

5.10 Lambert projection (viewed from true north)

jection is developed from the centre point of the geographic feature to be surveyed and charted. Using true north as an example, Figure 5.10 illustrates the Lambert projection.

NAVIGATION TERMINOLOGY

The terms shown in Table 5.1 are used with numerous navigation systems, including INS and RNAV; computed values are displayed on a control display unit (CDU) and/or primary flight instruments. These aircraft navigational terms are illustrated in Figure 5.11; all terms are referenced to true north.

NAVIGATION SYSTEMS DEVELOPMENT

meridians of the Mercator-projected chart are crossed at the same angle; the track followed is the rhumb line.

For aircraft navigation the Mercator projection might be satisfactory; however, if we want to navigate by great circle routes then we need true directions. An alternative projection used for aircraft navigation, and most popular maps and charts, is the Lambert azimuth equal-area projection. This projection was developed by Johann Heinrich Lambert (1728–1777) and is particularly useful in high latitudes. The pro-

This section provides a brief overview of the development of increasingly sophisticated navigation systems used on aircraft.

Directional gyro

The early aviators used visual aids to guide them along their route; these visual aids would have included rivers, roads, rail tracks, coastlines and so on. This type of navigation is not possible at high altitudes or

5.11 Navigation terminology

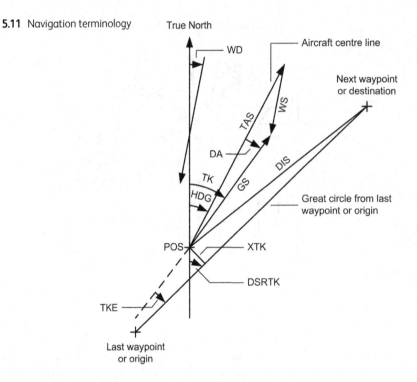

Table 5.1 Navigation terminology

Term	Abbreviation	Description
Cross track distance	XTK	Shortest distance between the present position and desired track
Desired track angle	DSRTK	Angle between north and the intended flight path of the aircraft
Distance	DIS	Great circle distance to the next waypoint or destination
Drift angle	DA	Angle between the aircraft's heading and ground track
Ground track angle	TK	Angle between north and the flight path of the aircraft
Heading	HDG	Horizontal angle measured clockwise between the aircraft's centre line (longitudinal axis) and a specified reference
Present position	POS	Latitude and longitude of the aircraft's position
Track angle error	TKE	Angle between the actual track and desired track (equates to the desired track angle minus the ground track angle)
Wind direction	WD	Angle between north and the wind vector
True airspeed	TAS	True airspeed measured in knots
Wind speed	WS	Measured in knots
Ground speed	GS	Measured in knots

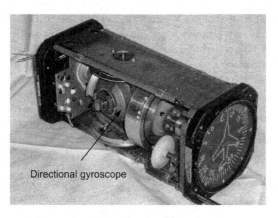

5.12 Directional gyro

in low visibility and so the earth's magnetic field was used as a reference, leading to the use of simple magnetic compasses in aircraft. We have seen that magnetic variation has to be taken into account for navigation; there are additional considerations to be addressed for compasses in aircraft. The earth's magnetic field around the aircraft will be affected by:

- The aircraft's own 'local' magnetic fields, e.g. those caused by electrical equipment
- Sections of the aircraft with high permeability causing the field to be distorted.

Magnetic compasses are also unreliable in the short term, i.e. during turning manoeuvres. Directional gyroscopes are reliable for azimuth guidance in the short term, but drift over longer time periods. A combined magnetic compass stabilized by a directional gyroscope (referred to as a gyromagnetic compass) can overcome these deficiencies. The directional gyro – Figure 5.12 – together with an airspeed indicator allows the crew to navigate by dead reckoning, i.e. estimating their position by extrapolating from a known position and then keeping note of the direction and distance travelled. In addition to directional references, aircraft also need an attitude reference for navigation, typically from a vertical gyroscope. Advances in sensor technology and digital electronics have led to combined attitude and heading reference systems (AHRS) based on laser gyros and microelectromechanical sensors.

Instrumentation errors inevitably lead to deviations between the aircraft's actual and calculated positions; these deviations accumulate over time. Crews therefore need to be able to confirm and update their position by means of a fixed ground-based reference, e.g. a radio navigation aid.

Radio navigation

Early airborne navigation systems using ground-based navigation aids consisted of a loop antenna in the aircraft tuned to amplitude modulated (AM) commercial radio broadcast stations transmitting in the low/medium-frequency (LF/MF) bands. Referring to Figure 5.13, pilots would know the location of the radio station (indeed, it would invariably have been

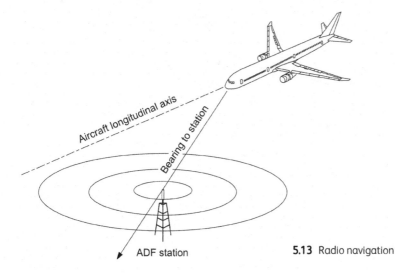

5.13 Radio navigation

located close to or even in the town/city that the crew wanted to fly to) and this provided a means of fixing a position. Although technology has moved on, these automatic direction finder (ADF) systems are still in use today. Operational problems are encountered using low-frequency (LF) and medium-frequency (MF) transmissions.

During the mid to late 1940s, it was evident to the aviation world that an accurate and reliable short-range navigation system was needed. Since radio communication systems based on very high frequency (VHF) were being successfully deployed, a decision was made to develop a radio navigation system based on VHF. This system became the VHF omnidirectional range (VOR) system, as seen in Figure 5.14, a system that is in widespread use throughout the world today. VOR is the basis of the current network of 'airways' that are used in navigation charts.

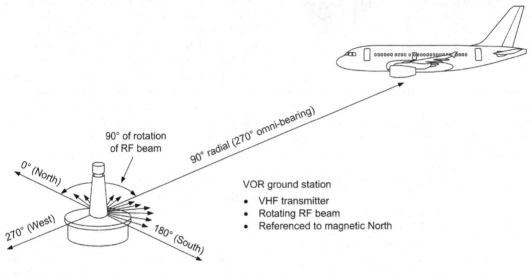

5.14 VOR radio navigation

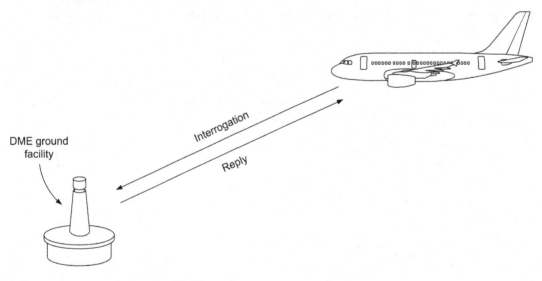

5.15 Distance measuring equipment (DME)

The advent of radar in the 1940s led to the development of a number of navigation aids, including distance-measuring equipment (DME). This is a short/medium-range navigation system, often used in conjunction with the VOR system to provide accurate navigation fixes. The DME system is based on secondary radar principles, as seen in Figure 5.15.

Navigation aids, such as automatic direction finder (ADF), VHF omnidirectional range (VOR) and distance-measuring equipment (DME), are used to define airways for en-route navigation, see Figure 5.16. They are also installed at airfields to assist with approaches to those airfields. These navigation aids cannot, however, be used for precision approaches and landings. The standard approach and landing system installed at airfields around the world is the instru-

ment landing system (ILS) – see Figure 5.17. The ILS uses a combination of VHF and UHF radio waves and has been in operation since 1946. There are a number of shortcomings with ILS; in 1978 the microwave landing system (MLS) was adopted as the long-term replacement. The system is based on the principle of time-referenced scanning beams and provides precision navigation guidance for approach and landing. MLS provides three-dimensional approach guidance, i.e. azimuth, elevation and range. The system provides multiple approach angles for both azimuth and elevation guidance. Despite the advantages of MLS, it has not yet been introduced on a worldwide basis for commercial aircraft. Military operators of MLS often use mobile equipment that can be deployed within hours.

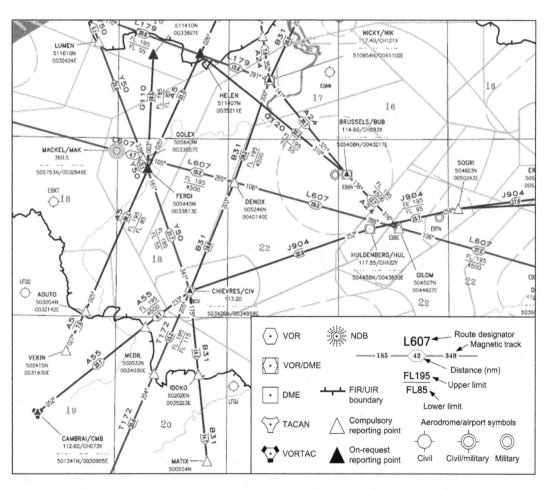

5.16 Airways defined by navigation aids

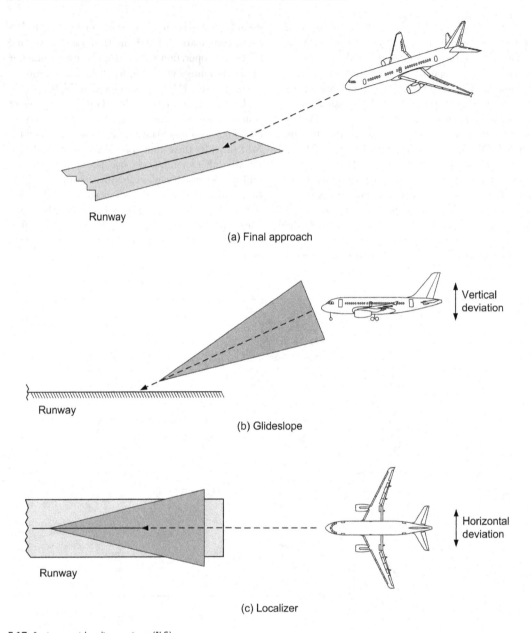

(a) Final approach

(b) Glideslope

(c) Localizer

5.17 Instrument landing system (ILS)

The aforementioned radio navigation aids have one disadvantage in that they are land based and only extend out beyond coastal regions. Long-range radio navigation systems based on hyperbolic navigation were introduced in the 1940s to provide for en-route operations over oceans and unpopulated areas. Several hyperbolic systems have been developed since, including Decca, Omega and Loran. The operational use of Omega and Decca navigation systems for aviation ceased in 1997 and 2000 respectively. Loran systems are still very much available today as stand-alone systems, in particular for marine applications.

The advent of computers, in particular the increasing capabilities of integrated circuits using digital

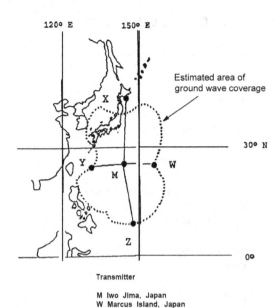

Transmitter

M Iwo Jima, Japan
W Marcus Island, Japan
X Hokkaido, Japan
Y Gesashi, Japan
Z Barrigada, Guam

5.18 Area navigation (RNAV)

techniques, has led to a number of advances in aircraft navigation. One example of this is the area navigation system (RNAV); this is a means of combining, or filtering, inputs from one or more navigation sensors and defining positions that are not necessarily co-located with ground-based navigation aids. Typical navigation sensor inputs to an RNAV system can be from external ground-based navigation aids such as VHF omnirange (VOR) and distance-measuring equipment (DME), see Figure 5.18.

Dead-reckoning systems

Dead-reckoning systems require no external inputs or references from ground stations. Doppler navigation systems were developed in the mid 1940s and introduced in the mid 1950s as a primary navigation system. Ground speed and drift can be determined using a fundamental scientific principle called Doppler shift. Being self-contained, the system can be used for long-distance navigation over oceans and undeveloped areas of the globe.

A major advance in aircraft navigation came with the introduction of the inertial navigation system (INS). This is an autonomous dead-reckoning system, i.e. it requires no external inputs or references from ground stations. The system was developed in the 1950s for use by the US military and subsequently the space programmes. Inertial navigation systems (INS) were introduced into commercial aircraft service during the early 1970s. The system is able to compute navigation data such as present position, distance to waypoint, heading, ground speed, wind speed, wind direction and so on. It does not need radio navigation inputs and it does not transmit radio frequencies. Again, being self-contained, the system can be used for long-distance navigation over oceans and undeveloped areas of the globe.

Global navigation satellite systems

Navigation by reference to the stars and planets has been employed since ancient times; commercial aircraft used to have periscopes to take celestial fixes for long-distance navigation. An artificial constellation of navigation aids was initiated in 1973 and referred to as Navstar (navigation system with timing and ranging). The global positioning system (GPS) was developed for use by the US military; the first satellite was launched in 1978 and the full constellation was in place and operating by 1994. GPS is now widely available for use by many applications including aircraft navigation; the system calculates the aircraft position by triangulating the distances from a number of satellites – see Figure 5.19. Galileo is the European global navigation satellite system (GNSS) currently being introduced. It provides a high-precision positioning system upon which European nations can rely, independently from the Russian GLONASS, US GPS and Chinese Compass systems.

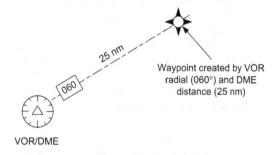

5.19 Satellite navigation

TEST YOUR UNDERSTANDING 5.1

The nautical mile is directly linked to the geometry of the earth. How is a nautical mile defined?

TEST YOUR UNDERSTANDING 5.2

Explain the difference between dead reckoning and position fixing.

TEST YOUR UNDERSTANDING 5.3

For a given airspeed, explain how tailwinds and headwinds affect groundspeed.

NAVIGATION SYSTEMS SUMMARY

Navigation systems for aircraft have evolved with the nature and role of the aircraft itself. These individual systems are described in detail in the following chapters. Each system has been developed to meet specific requirements within the available technology and cost boundaries. Whatever the requirement, all navigation systems are concerned with several key factors:

- Accuracy: conformance between calculated and actual position of the aircraft
- Integrity: ability of a system to provide timely warnings of system degradation
- Availability: ability of a system to provide the required function and performance
- Continuity: probability that the system will be available to the user
- Coverage: geographic area where each of the above are satisfied at the same time.

A full account of aircraft navigation is given in an accompanying title in this book series, *Aircraft Communications and Navigation Systems*.

MULTIPLE-CHOICE QUESTIONS

1. Longitude referenced to the prime meridian extends:

   (a) North or south up to 180°
   (b) East or west up to 180°
   (c) East or west up to 90°

2. Latitude is the angular reference:

   (a) North or south of the equator
   (b) East or west of the prime meridian
   (c) North or south of the prime meridian

3. The distance between lines of longitude converge at the:

   (a) Poles
   (b) Equator
   (c) Great circle

4. Lines of latitude are always:

   (a) Converging
   (b) Parallel
   (c) The same length

5. Degrees of latitude can be divided by 60, giving the unit of:

   (a) Longitude
   (b) Minutes
   (c) Seconds

6. The location of magnetic north is approximately:

   (a) 80° latitude and 110° longitude, east of the prime meridian
   (b) 80° longitude and 110° latitude, west of the prime meridian
   (c) 80° latitude and 110° longitude, west of the prime meridian

7. One minute of arc of a great circle defines a:

   (a) Nautical mile
   (b) Kilometre
   (c) Knot

8. The angular difference between magnetic north and true north is called the:

   (a) Magnetic variation
   (b) Great circle
   (c) Prime meridian

9. Mercator projections produce parallel lines of:

   (a) The earth's magnetic field
   (b) Longitude
   (c) Great circle routes

10. With respect to the polar radius, the equatorial radius of the earth is:

    (a) Equal
    (b) Larger
    (c) Smaller

6   Control systems

Control systems are designed in many forms and are utilized throughout a range of aircraft applications. A specific input, such as moving a lever or joystick, causes a specific output, such as feeding current to an electric motor that in turn operates a hydraulic actuator that moves, e.g. the elevator of the aircraft. At the same time, the position of the elevator is detected and fed back to the pitch attitude controller, so that small adjustments can continually be made to maintain the desired attitude and altitude.

Control systems invariably comprise a number of elements, components or sub-systems that are connected together in a particular way. The individual elements of a control system interact together to satisfy a particular functional requirement, such as modifying the position of an aircraft's control surfaces.

ELEMENTARY CONTROL

The simplest control systems can be illustrated by electrical circuits; in Figure 6.1, these simple circuits are used to control (a) a lightbulb and (b) a motor. Closing the switch completes the circuit and allows electrical energy to flow from the battery to the load, where the electrical energy is converted into the desired function, e.g. light or movement. Opening the switch stops the supply of energy to the load. The energy required to open or close the switch has no relationship to the amount of energy being controlled. In the example given, the controlling energy is mechanical; the controlled energy is electrical; the resulting energy is (a) light and (b) kinetic. (Heat is often a byproduct.)

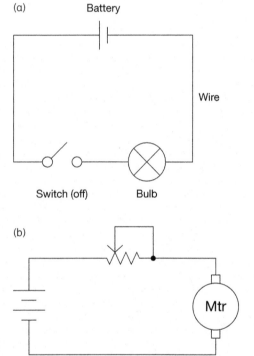

6.1 Simple control systems: simple electrical circuits – (a) lightbulb (b) motor

6.2 Water turbine control system

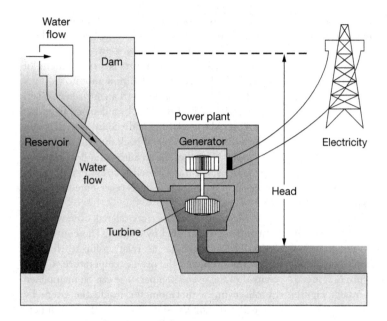

Another example of control systems is illustrated in Figure 6.2, which depicts a water turbine system. Water is contained in a header tank, or some other form of reservoir. Opening the valve allows water to flow through the turbine, which causes the turbine to rotate; the water then exits the system via the outlet. The potential energy of the water is converted into kinetic energy in the turbine; the two are controlled by the valve. As with the electrical circuits, the work required to open or close the valve is independent of the rate of energy conversion. One important difference between the water turbine example and the given electrical circuits is that there are an infinite number of positions that the water control valve can be adjusted to. The rate of energy conversion can be controlled to any desired value between zero and maximum.

The electrical equivalent of this variable control can be illustrated by replacing the on/off switch with a rheostat – see Figure 6.3. When the sliding contact of the rheostat is in position A, the electrical circuit is open, no current flows and the motor does not turn. When the sliding contact of the rheostat is in position B, maximum current flows in the circuit and the motor turns at its fastest speed. At any position of the sliding contact of the rheostat between positions A and B there will be an interim motor speed proportional to circuit current, which is determined by the slider position. As with the previous examples, the amount

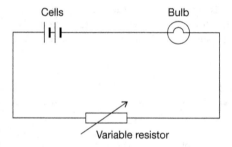

6.3 Rheostat control system

of energy required to move the slider has no relationship to the amount of energy supplied by the battery, or given by the motor.

The final example is illustrated with a simple audio system – see Figure 6.4. When sound is sensed by the microphone, its diaphragm vibrates and a vari-

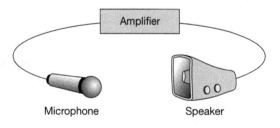

6.4 Simple audio control system

ation of current is created in the circuit. This current variation is amplified and the output fed to a speaker. Depending on the circuit arrangement, amplifier design and so on, the resulting sound energy from the speaker is higher than the input to the microphone.

CONTROL SYSTEMS

When the control function is independent of the obtained result, the system is called an open-cycle system. In an open-loop control, the value of the input variable is set to a given value in the expectation that the output will reach the desired value. In such a system there is no automatic comparison of the actual output value with the desired output value in order to compensate for any differences. A simple open-loop system schematic is represented in Figure 6.5; it can be illustrated with the analogy of a water tank, Figure 6.6.

Water flows into the tank from the supply when the valve is opened. When the water in the tank reaches the desired level, the valve is turned off. The actions of opening and closing the valve, and deciding

on the required water level, are both manual operations; there is no automatic link between these actions, i.e. it is an open-loop system.

Now consider the water tank arrangement of Figure 6.7; the water level is now monitored by a sensor that transmits a signal to a controller. When water is taken out of the tank by a demand from the system user, the level drops and a control signal is sent to the valve. Water is piped back into the tank from the supply. Now there is an automatic link between these actions, i.e. it is a closed-loop system. A simple closed-loop system schematic is represented in Figure 6.8.

To illustrate this further, consider an oven with one heat setting, controlled by a timer. The controlling action of the timer has no relationship to the heat inside the oven, and the oven's temperature. Closer control of the oven's temperature can be maintained by observing a thermometer inside the oven; by manually turning the system on or off, the temperature can be regulated. The switch is turned on by a person when the temperature falls below the desired value, and off when the temperature goes above the desired value.

In this example, the person actually closes the loop on an intermittent basis. In effect, the oven relies on human intervention in order to ensure consistency of the food produced. Clearly, open-loop control has some significant disadvantages for more complex systems. What is required is some means of closing the loop in order to make a continuous automatic

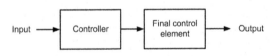

6.5 Open-loop control system schematic

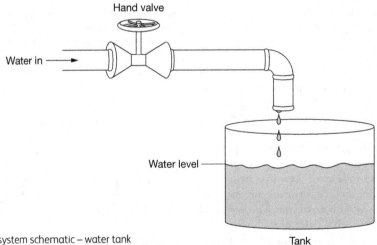

6.6 Open-loop system schematic – water tank

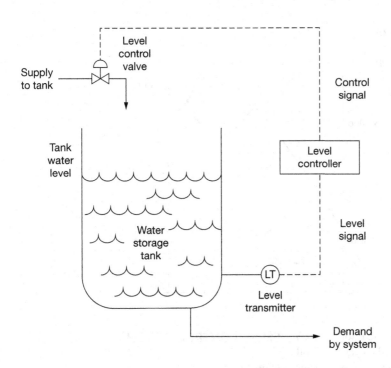

6.7 Closed-loop system schematic – water tank

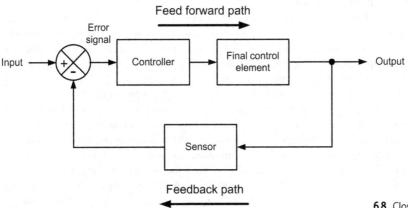

6.8 Closed-loop system schematic

comparison of the actual value of the output compared with the setting of the input control variable.

Control of the oven temperature could be automated by substituting the person and thermometer with a bi-metallic switch or thermostat, as in Figure 6.9. When the oven is switched on, but below temperature, the thermostat is closed and the heating element starts to warm up. When the desired temperature is achieved, the thermostat opens and the heating element is disconnected from the supply. The oven temperature is automatically regulated by the

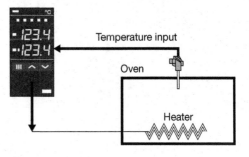

6.9 Oven temperature control system

ON/OFF Control

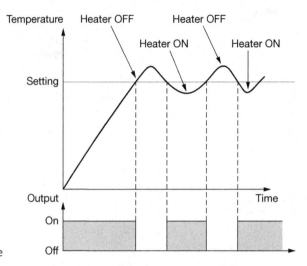

6.10 Temperature control profile

thermostat, i.e. there is no requirement for a person to do anything other than to turn the oven on or off as required. The relationship between the heater on/off control and temperature profiles over a given time period is shown in Figure 6.10.

There are many advantages of closed-loop systems. Some systems use a very large number of input variables and it may be difficult or impossible for a human operator to keep track of them. Control processes can be extremely complex and there may be significant interaction between the input variables. Some systems may have to respond very quickly to changes in variables (human reaction times may not be fast enough). Many systems require a very high degree of precision; human operators may not be able to work to a sufficiently high degree of accuracy.

> **KEY POINT**
>
> When a control system requires a person to monitor the input/output, it is termed an open-loop system. If the input/output is controlled automatically, it is termed closed loop.

SERVO CONTROL SYSTEMS

Overview

Servomechanisms, or servo control systems, are utilized in many aircraft applications to automatically control the operation of systems. The servo control system is based on changes of a control input causing corresponding output changes. Transducers are devices used to convert the desired parameter, e.g. pressure, temperature, displacement and so on, into electrical energy. The transducer's output signal is amplified, applied to a motor that is used to control an output load. A practical closed-loop control system used to control the speed of an output shaft is illustrated in Figure 6.11. The desired input speed is set by the operator by selecting a position on a potentiometer. An amplifier compares the selected speed and compares this to the feedback from the output load's speed. Difference signals (termed errors) from the output are fed back to the error detector so that the necessary corrections can be made to reduce and ultimately eliminate the difference, or error, between input and output.

Closed-loop systems are used to convert small input signals (typically from a transducer) into larger outputs (typically via an actuator); the relationship between the input and the output is proportional and in accordance with a predefined gain, or control laws. The practical application of servo control systems

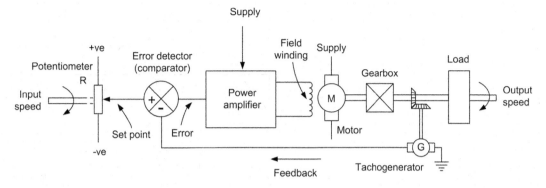

6.11 Closed-loop speed control system:

is covered in subsequent automatic pilot chapters. In the example described, the servo motor feedback is provided by a position sensor; some servo control motors give feedback as motor speed or acceleration.

System response

In a perfect theoretical system, the output will respond instantaneously to a change in the input value (set point). The theoretical servo amplifier must be able to provide an infinite acceleration. There will be no delay when changing from one value to another and no time required for the output to settle to its final value. In practice, real-world systems take time to reach their final state. The practical servo system has a critically damped response, necessary to prevent oscillation. Very sudden changes in output may, in some cases, be undesirable. Furthermore, friction and inertia are present in many systems.

The inertia of the load will effectively limit the acceleration of the output motor. Furthermore, as the output speed reaches the desired value, the inertia present will keep the speed increasing despite the reduction in voltage applied to the motor. Thus, the output shaft speed will overshoot the desired value, before eventually falling back to the required value. There are two types of input to a servo system, ramp inputs and step inputs – see Figure 6.12. Ramp inputs gradually change at a given speed; step inputs change instantaneously.

A step input occurs when the input is changed very quickly, e.g. in the case of the speed input control system, the control input is instantaneously set to a new position. The system's inertia through the motor

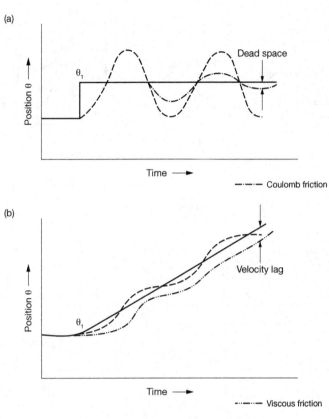

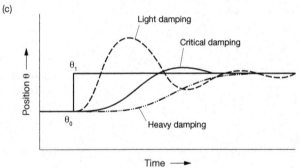

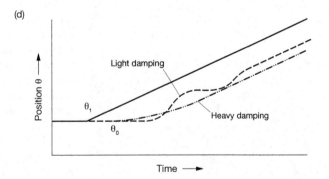

6.12 Servo systems inputs: (a) step,
(b) ramp, (c) damping of step input,
(d) damping of ramp input

and gears cannot respond instantaneously, so there is a large error signal. The load is accelerated to the new position, thereby reducing the error to zero. The load is now responding at a constant rate and will eventually overshoot the required speed. The overshoot causes an error signal in the opposite direction, and the output speed is now reduced towards the desired speed. If frictional losses are zero, a continuous oscillation is created, with a constant dead space in the system. In practical systems, there will be some friction, and the system's dead space will eventually reduce to zero.

A ramp input occurs with a smooth rate of change of the controlling input. The error signal is initially small; the output accelerates slowly, but still lags behind the input. There is an oscillatory response as shown, with a corresponding velocity lag.

Friction is a force phenomenon which opposes the relative movement between two contact surfaces. It is a physical phenomenon, expressed in qualitative terms as the force exerted by either of the contacting bodies tending to oppose relative tangential displacement of the other. Friction is commonly modelled as a linear combination of:

• Coulomb friction
• Stiction
• Viscous friction

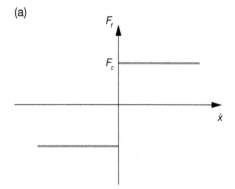

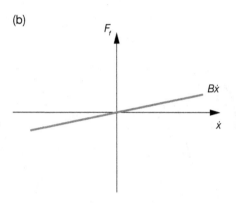

6.13 (a) Coulomb friction, (b) viscous friction

Coulomb friction, named after the French physicist Charles-Augustin de Coulomb (1736–1806), is a constant friction contribution, i.e. it is not dependent on velocity. In servo systems, steady-state errors and tracking errors are mainly caused by static friction (stiction), which depends on the velocity's direction, and the viscous friction that increases the damping of a system. Although the Coulomb friction model simplifies the frictional phenomena, it is widely used in servo control systems if dynamic effects are not a consideration. The Coulomb friction force (F_C) is of constant magnitude, acting in the direction opposite to motion, as in Figure 6.13(a).

One of the biggest considerations for the Coulomb model is that it does not account for the zero velocity condition, hence the properties of motion at starting or zero velocity crossing, i.e. static and rising static friction, or stiction.

Viscous friction is proportional to the relative movement, or sliding velocity. Here, the friction is considered to be proportional to the velocity, and

is expressed as a function of a viscous friction coefficient, B, multiplied by the velocity, as seen in Figure 6.13(b). The force needed to maintain a constant velocity is the same magnitude as the viscous friction force, but in the opposite direction.

In some extreme cases the oscillation which occurs when the output value cycles continuously above and below the required value may be continuous. The oscillatory component can be reduced (or eliminated) by artificially slowing down the response of the system; this is known as damping. Oscillations in a control system are undesirable; damping is used to reduce the overshoots. Inherent friction will provide light damping, and maximum overshoot. Heavy damping eliminates the overshoot, but means that the system is unresponsive, or sluggish. Critical damping gives one small overshoot and allows the system to provide the optimum response.

Increasing the gain of the system will have the effect of increasing the acceleration, but this, in turn, will also produce a correspondingly greater value of overshoot.

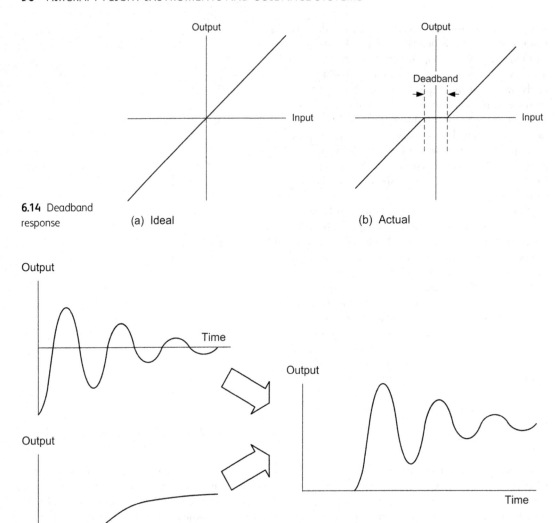

6.14 Deadband response

(a) Ideal   (b) Actual

6.15 Growth curve and damped oscillation

Conversely, decreasing the gain will reduce the overshoot but at the expense of slowing down the response.

Deadband refers to the inability of a control system to respond to a small change in the input; in other words, the input changes but the output does not. Deadband is illustrated by the ideal and actual system response shown in Figure 6.14. Deadband can be reduced by increasing the gain present within the system but, as stated earlier, this may have other undesirable effects.

KEY POINT

Servo system feedback has to be negative to minimize oscillations. Positive feedback would cause the servo system to become unstable.

The ideal response consists of a step input function, whilst the actual response builds up slowly and shows a certain amount of overshoot. The response of a control system generally has two basic components: an exponential growth curve and a damped oscillation – see Figure 6.15.

MULTIPLE-CHOICE QUESTIONS

1. An open-loop control system is one in which:

   (a) No feedback is applied
   (b) Positive feedback is applied
   (c) Negative feedback is applied

2. The range of outputs close to the zero point that a control system is unable to respond to is referred to as:

   (a) Hunting
   (b) Overshoot
   (c) Deadband

3. A closed-loop control system is one in which:

   (a) No feedback is applied
   (b) Negative feedback is applied
   (c) Positive feedback is applied

4. The gain of a servo control system determines:

   (a) How fast the servo motor tries to reduce the error
   (b) Damping
   (c) If it is open or closed loop

5. Servo system feedback has to be negative to minimize:

   (a) Deadband
   (b) Gain
   (c) Oscillations

6. Overshoot in a control system can be reduced by:

   (a) Reducing the damping
   (b) Increasing the gain
   (c) Increasing the damping

7. Decreasing the gain of a servo system will:

   (a) Reduce the overshoot and slow down the response
   (b) Reduce the overshoot and speed up the response
   (c) Increase the overshoot but slow down the response

8. The optimum value of damping is that which allows:

   (a) No overshooting
   (b) Increasing overshooting
   (c) One small overshoot

9. Ramp inputs are characterized by:

   (a) Gradually changing at a given speed
   (b) Changing instantaneously
   (c) High gain servo systems

10. The output of a control system continuously oscillating above and below the required value is known as:

    (a) Deadband
    (b) Hunting
    (c) Overshoot

7 Aeroplane aerodynamics

This chapter serves as an introduction to aerodynamics and theory of flight for subsonic aeroplanes to underpin the study of autopilots and flight guidance systems. By EASA's definition, 'aeroplane' means an engine-driven fixed-wing aircraft heavier than air that is supported in flight by the dynamic reaction of the air against its wings. The study of elementary flight theory in this chapter will be of interest for all aircraft engineers, no matter what their trade specialization. In particular, there is a need for engineers to understand how aircraft produce lift and how they are controlled and stabilized for flight. This knowledge will then assist engineers with their future understanding of autopilot and control systems and the importance of the design features that are needed to stabilize aircraft during all phases of flight. The requisite knowledge needed for the successful study and completion of basic aerodynamics, as laid down in Module 8 of EASA's Part 66 syllabus, is addressed by another title in this book series, *Aircraft Engineering Principles*. Full coverage of aircraft flight control, control devices and high-speed flight theory will be found in another book in the series, *Aircraft Aerodynamics, Flight Control and Airframe Structures*.

Our study of aerodynamics is based on the important topics covered previously in Chapter 1, including the nature and purpose of the international standard atmosphere (ISA). This chapter also draws on knowledge of fluids in motion for the effects of airflow over aircraft and the underlying physical principles that account for the creation of aircraft lift and drag.

STATIC AND DYNAMIC PRESSURE

Overview

Fluid in steady motion has both static pressure energy and dynamic pressure energy (kinetic energy) due to the motion. Bernoulli's equation showed that for an ideal fluid, the total energy in a steady streamline flow remains constant.

> **KEY POINT**
>
> Bernoulli's equation: static pressure energy + dynamic (kinetic) energy = Constant total energy.

> **KEY POINT**
>
> Bernoulli's equation expressed in mathematical terms: $p + \frac{1}{2}\rho v^2 = C$

With respect to aerodynamics, dynamic pressure is dependent on the density of the air (treated as an ideal fluid) and the velocity of the air. Thus, with increase in altitude there is a drop in density and the dynamic pressure acting on the aircraft as a result of the airflow will also drop with increase in altitude. The static pressure of the air also drops with increase in altitude.

Venturi tube

An important application of Bernoulli's equation for aerodynamics is the Venturi tube, named after Giovanni Battista Venturi (1746–1822), an Italian physicist. This is based on studies of fluids in motion; from Bernoulli's equation above, use is made of the fact that to maintain equality, an increase in velocity will mean a decrease in static pressure, or alternatively, a decrease in velocity will mean an increase in static pressure.

Flow through a Venturi tube, as shown in Figure 7.1, illustrates the principles of Bernoulli's theorem; the tube gradually converges to a throat, and then diverges even more gradually. If measurements are taken at the throat, a decrease in pressure will be observed. Now according to Bernoulli's equation a reduction in static pressure must be accompanied by an increase in dynamic pressure if the relationship is to remain constant. The increase in dynamic pressure is achieved by an increase in the velocity of the fluid as it reaches the throat. The effectiveness of the Venturi tube as a means of causing a decrease in pressure below that of the atmosphere depends very much on its geometric profile.

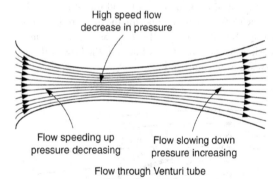

High speed flow
decrease in pressure

Flow speeding up
pressure decreasing

Flow slowing down
pressure increasing

Flow through Venturi tube

7.1 Venturi tube

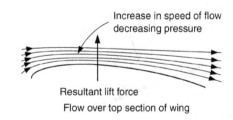

Increase in speed of flow
decreasing pressure

Resultant lift force
Flow over top section of wing

7.2 Venturi tube/top surface of wing analogy

The Venturi tube provides us with the key to the generation of lift. Imagine that the bottom cross-section of the tube is the top part of an aircraft wing, shown in cross-section in Figure 7.2. Then the increase in velocity of flow over the wing causes a corresponding reduction in pressure, below atmospheric pressure. It is this reduction in pressure which provides the lift force perpendicular to the top surface of the wing, and due to the shape of the lower wing cross-section a slight increase in pressure is achieved, which also provides a component of lift.

SUBSONIC AIRFLOW

Flow over a flat plate

When a body is moved through the air, or any fluid that has viscosity, such as water, there is a resistance produced which tends to oppose the body. For example, if you are driving in an open-topped car, there is a resistance from the air acting in the opposite direction to the motion of the car. This air resistance can be felt on your face or hands as you travel. In the aeronautical world, this air resistance is known as drag. It is undesirable for obvious reasons. For example, aircraft engine power is required to overcome this air resistance and unwanted heat is generated by friction as the air flows over the aircraft hull during flight.

We consider the effect of air resistance by studying the behaviour of airflow over a flat plate. If a flat plate is placed edge-on to the relative airflow, as in Figure 7.3, then there is little or no alteration to the smooth passage of air over it. The airflow over the plate is said to be streamline, or laminar. On the other hand, if the plate is offered into the airflow at some angle of inclination to it, it will experience a reaction that tends to both lift it and drag it back. This is the same effect that you can feel on your hand when placed into the airflow as you are travelling, e.g. in the open-topped car mentioned earlier. The amount of reaction depends upon the speed and angle of inclination between the flat plate and relative airflow.

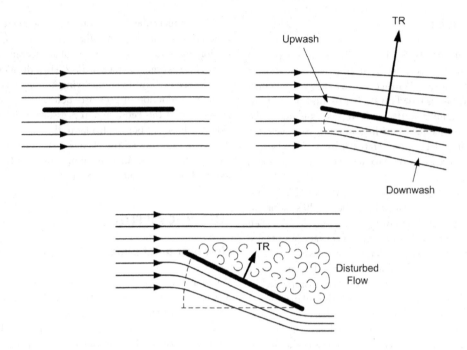

7.3 Flat plate and relative airflow

> **KEY POINT**
>
> Streamline or laminar flow is where fluid particles move in an orderly manner and retain their relative positions in successive cross-sections. The flow pattern maintains the profile of the body that it flows over.

as shown in Figure 7.4. One at right angles to the relative airflow, known as lift, and the other parallel to the relative airflow, opposing the motion, known as drag. The drag force is the same as that mentioned earlier, which caused a resistance to the flow of the air stream over your hand.

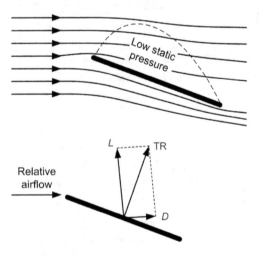

7.4 Reaction components of a flat plate in airflow

As can be seen, when the flat plate is inclined at an angle to the relative airflow, the streamlines are disturbed. An upwash is created at the front edge of the plate, causing the air to flow through a more constricted area, in a similar manner to flow through the throat of a Venturi tube. The net result is that as the air flows through this restricted area, it speeds up. This in turn causes a drop in static pressure above the plate (Bernoulli) when compared with the static pressure beneath it, resulting in a net upward reaction. After passing the plate, there is a resulting downwash of the air stream.

The total reaction on the plate caused by it disturbing the relative airflow has two reactive components,

Streamline flow, laminar flow and turbulent flow

Streamline flow is flow in which the particles of the fluid move in an orderly manner and retain the same relative positions in successive cross sections, maintaining the shape of the body over which it is flowing. This streamline flow is illustrated in Figure 7.5, where it can be seen that the successive cross sections are represented by lines that run parallel to one another hugging the shape of the body around which the fluid is flowing. Laminar flow can be depicted as the smooth parallel layers of air flowing over the surface of a body in motion. Turbulent flow occurs when the particles of fluid move in a disorderly manner, occupying different relative positions in successive cross sections, as seen in Figure 7.6.

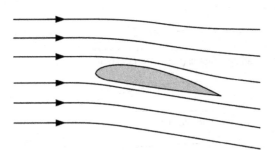

7.5 Streamline flow

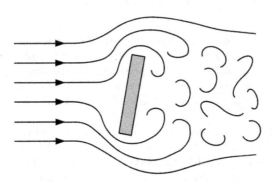

7.6 Turbulent flow

AEROFOILS

Overview

An aerofoil profile is designed to obtain a desirable reaction from the air through which it moves. It converts the physical properties of the surrounding air into a useful force that produces lift for flight. The cross section of a conventional aircraft wing is an aerofoil section, where the top surface usually has greater curvature than the bottom surface.

The air approaching an aerofoil section, as shown in Figure 7.7, is split as it passes around the aerofoil. Over the top surface it will speed up, because it must reach the trailing edge of the aerofoil at the same time as the air that flows underneath the section. In doing so, there must be a decrease in the pressure of the airflow over the top surface that results from its increase in velocity (Bernoulli).

The top half of the aerofoil surface shown in Figure 7.8 can be considered as the bottom half of a Venturi tube. As the airflow passes through the restriction, there will be a corresponding increase in the speed of the airflow, in a similar manner to that created by the upwash for a flat plate mentioned earlier. Then again, from Bernoulli's equation, this increase in speed causes a corresponding decrease in static pressure, as shown, which over the surface of the aerofoil will create the desired lift force.

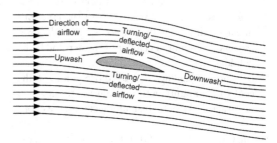

7.7 Air approaching an aerofoil section

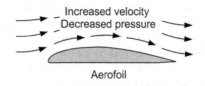

7.8 Top half of an aerofoil surface

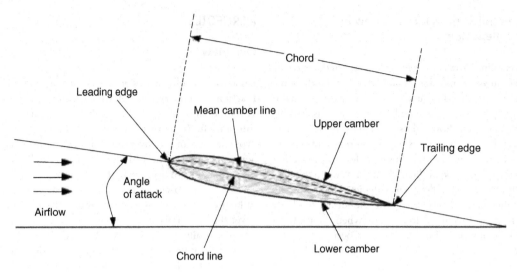

7.9 Aerofoil terminology

Aerofoil terminology

Terms and definitions related to aerofoils that will be used frequently throughout this chapter are illustrated in Figure 7.9.

- Camber refers to the upper and lower curved surfaces of the aerofoil section.
- Mean camber line is halfway between the upper and lower cambers.
- Chord line joins the centres of curvatures of the leading and trailing edges. (Note that the chord line may fall outside the aerofoil section, depending on the amount of camber of the aerofoil being considered.)
- Leading and trailing edge are those points on the centre of curvature of the leading and trailing part of the aerofoil section that intersect with the chord line.
- Angle of incidence (AOI) is the angle between the relative airflow and the longitudinal axis of the aircraft. It is a built-in feature of the aircraft and is referred to as the 'rigging angle'. On conventional aircraft, the AOI is designed to minimize drag during cruise, thus maximizing fuel efficiency.
- Angle of attack (AOA) is the angle between the chord line and the relative airflow. This will vary, depending on the longitudinal attitude of the aircraft, with respect to the relative airflow, as will be described later.

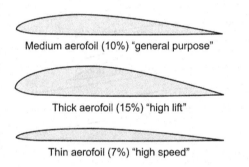

7.10   Selection of aerofoil sections

- Thickness/chord ratio is the ratio of the maximum thickness of the aerofoil section to its chord length, normally expressed as a percentage. It is sometimes referred to as the fineness ratio and is a measure of the aerodynamic thickness of the aerofoil.

The aerofoil shape is also defined in terms of its thickness/chord ratio. The aircraft designer chooses that shape which best meets the aerodynamic requirements of the aircraft. A selection of aerofoil sections is shown in Figure 7.10.

Aerofoil efficiency

The efficiency of an aerofoil is measured using the lift-to-drag (L/D) ratio. As you will see when we study lift and drag, this ratio varies with changes in the AOA, reaching a maximum at one particular AOA. For conventional aircraft using wings as their main source of lift, maximum L/D is found to be around 3 or 4 degrees. Thus, we set the wings at an incidence angle of 3 or 4 degrees when the aircraft is flying straight. A typical lift/drag curve is shown in Figure 7.11, where the AOA for normal flight will vary from 0 degrees to around 15 or 16 degrees, at which point the aerofoil will stall.

Research has shown that the most efficient aerofoil sections for general use have their maximum thickness occurring around a third of the way back from the leading edge of the wing. It is thus the shape of the aerofoil section that determines the AOA at which the wing is most efficient and the degree of this efficiency. High-lift devices, such as slats, leading-edge flaps and trailing-edge flaps, alter the shape of the aerofoil section in such a way as to increase lift.

However, the penalty for this increase in lift is an increase in drag, which has the overall effect of reducing the L/D ratio.

Effects on airflow with changing AOA

The point on the chord line through which the resultant lift force acts is known as the centre of pressure (CP). The CP through which the resultant force or total reaction acts is shown in Figure 7.12. As the aerofoil varies its AOA, the air flow and pressure changes around the surface are affected. The CP will move forward along the chord line, as shown in Figure 7.13. For all positive AOA, the CP moves forward as the aircraft attitude or pitch angle is increased, until the stall angle is reached, when it suddenly moves backwards along the chord line. Note that the aircraft pitch angle should not be confused with the AOA. This

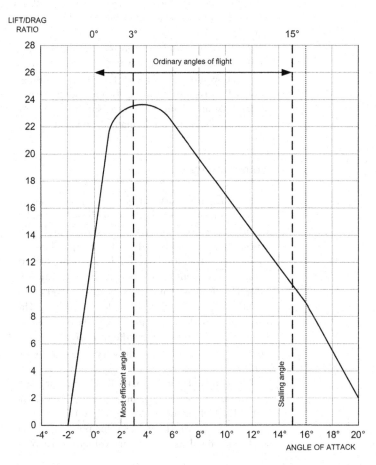

7.11 Typical lift/drag curve

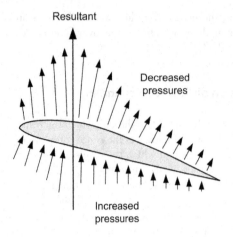

7.12 Centre of Pressure (CP) and resultant force

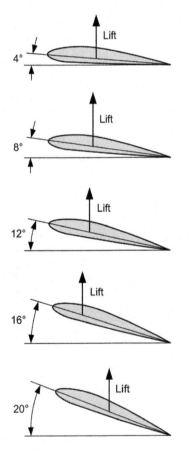

7.13 CP movement along the chord line

is because the relative airflow will change direction in flight in relation to the pitch angle of the aircraft.

Aerofoil stall

When the AOA of the aerofoil section is increased gradually towards a positive angle, the lift component increases rapidly up to a certain point, and then suddenly begins to drop off. When the AOA increases to a point of maximum lift, the stall point is reached – this is known as the critical angle or stall angle. When the stall angle is approached, the air ceases to flow smoothly over the top surface of the aerofoil and it begins to break away, as in Figure 7.15, creating turbulence.

At the critical angle, the pressure gradient is large enough to actually push a flow up the wing against the normal flow direction. This has the effect of causing a reverse flow region below the normal boundary layer, which separates from the aerofoil surface. When the aerofoil stalls, there is a dramatic drop in lift.

Boundary layer

When an aerofoil has air flowing past it, the air molecules in contact with the aerofoil surface tend to be brought to rest by friction. The next molecular layer of the air tends to bind to the first layer by molecular attraction, but tends to shear slightly, creating movement with respect to the first stationary layer. This process continues as successive layers shear slightly, relative to the layer underneath them. This produces a gradual increase in velocity of each successive layer of the air until the free stream relative velocity is reached some distance away from the aerofoil.

Referring to Figure 7.16, the fixed-boundary layer represents the skin of an aircraft wing, where the initial layer of air molecules has come to rest on its surface. The moving boundary is the point where the air has regained its free stream velocity relative to

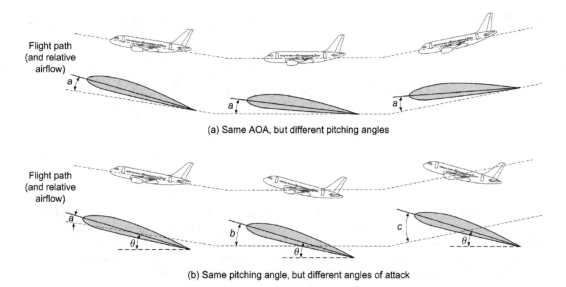

(a) Same AOA, but different pitching angles

(b) Same pitching angle, but different angles of attack

7.14 Pitch angle and AOA

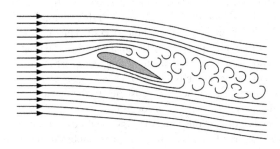

7.15 Approach to the stall angle

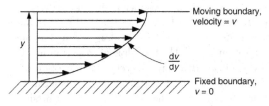

7.16 Boundary layer

the wing. The region between the fixed boundary and moving boundary, where this shearing takes place, is known as the boundary layer.

For an aircraft, subject to laminar flow over its wing section, the thickness of the boundary layer is seldom more than 1mm. The thinner the boundary layer, the less the drag and the greater the efficiency of the lift-producing surface. Since friction reduces the energy of the air flowing over an aircraft wing, it is important to keep wing surfaces and other lift-producing devices as clean and as smooth as possible. This will ensure that energy losses in the air close to the boundary are minimized and efficient laminar flow is maintained for as long as possible.

Irrespective of the smoothness and condition of the lift-producing surface, as the airflow continues back from the leading edge, friction forces in the boundary layer continue to use up the energy of the airstream, gradually slowing it down. This results in an increase in thickness of the laminar boundary layer with increase in distance from the leading edge – see Figure 7.17. Some distance back from the leading edge, the laminar flow begins an oscillatory disturbance, which is unstable. An eddying starts to occur, which grows larger and more severe until the smooth laminar flow is destroyed. Thus, a transition takes place in which laminar flow decays into turbulent boundary-layer flow.

Boundary-layer control devices provide an additional means of increasing the lift produced across an aerofoil section. In effect, all these devices are designed to increase the energy of the air flowing in the boundary layer, thus reducing the rate of boundary-layer separation, from the upper surface of

7.17 Boundary-layer separation

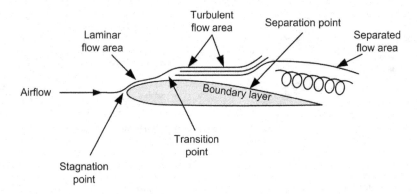

the aerofoil. At high AOA, the propensity for boundary-layer separation increases as the airflow over the upper surface tends to separate and stagnate.

Boundary-layer control devices include leading-edge slats, trailing and leading-edge flaps for high-lift applications, as mentioned previously. With slats the higher-pressure air from beneath the aerofoil section is 'sucked' over the aerofoil upper surface, through the slot created by deploying the slat. This high velocity air re-energizes the stagnant boundary-layer air, moving the transition point further back and increasing lift.

One fixed device that is used for boundary-layer control is the vortex generator. These generators are literally small metal plates that are fixed obliquely to the upper surface of the wing or other lift-producing surface and effectively create a row of convergent ducts close to the surface. These accelerate the airflow and provide higher velocity air to re-energize the boundary layer.

Many other devices exist or are being developed to control the boundary layer, which include blown air, suction devices and use of smart devices. More of this subject will be covered in another book in this series.

We have already discovered that the lift generated by an aerofoil surface is dependent on the shape of the aerofoil and its AOA to the relative airflow. Thus, the magnitude of the negative pressure distributed over the top surface of the wing is dependent on the wing camber and the wing AOA.

The shape of the aerofoil may be represented by a shape coefficient, which alters with AOA; this is known as the lift coefficient (C_L), which may be found experimentally for differing aerofoil sections. A corresponding drag coefficient (C_D) may also be determined experimentally or analytically, if the lift

coefficient is known. In addition to the lift and drag coefficients, the pitching moment, or the tendency of the aerofoil to revolve about its centre of gravity (CG), can also be determined experimentally.

DRAG

To complete our understanding of lift and drag, we need to define the different types of drag that affect the performance of the whole aircraft. This section describes all types of drag that make up the total drag acting on an aircraft; this total drag comprises many elements as shown in Figure 7.18.

Total drag is the total resistance to the motion of the aircraft as it passes through the air; it is the sum total of the various drag forces acting on the aircraft. These drag forces may be divided into subsonic drag and supersonic drag. Although supersonic drag is shown for completeness, it is addressed in more detail in another book in the series. Subsonic-flight drag may be divided into two major categories: these are profile drag and induced drag. Profile drag is further subdivided into skin-friction drag, form drag and interference drag. The total drag of an aircraft may be divided in another way. Whereby, the drag of the lift-producing surfaces, lift-dependent drag, is separated from those parts of the aircraft that do not produce lift. This non-lift-dependent drag is often known as parasite drag, it is the drag which results from the wing to fuselage shape and frictional resistance.

Skin-friction drag

Skin-friction drag results from the frictional forces that exist between a body and the air through which

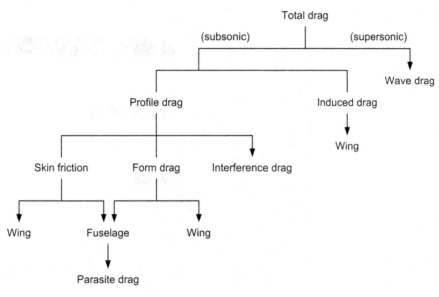

7.18 Elements of total drag

it is moving. The magnitude of the skin-friction drag depends on the surface area of the aircraft since the whole surface area of the aircraft experiences skin-friction drag. Surface roughness is another factor – the rougher the surface, the greater the skin-friction drag. As is the state of boundary-layer airflow, i.e. whether it is laminar or turbulent.

> **KEY POINT**
>
> External aircraft surfaces will affect drag; this can be reduced with polished metal and/or a good paint finish.

Form drag

Form drag is that part of the air resistance created by the shape of the body, subject to the airflow. Those shapes which encourage the airflow to separate from their surface create eddies and the streamline flow is disturbed. The turbulent wake that is formed increases drag. Form drag can be reduced by streamlining the aircraft in such a way as to reduce the drag resistance to a minimum. A definite relationship exists between the length and thickness of a streamlined body; this is known as the fineness ratio.

Streamlining shapes reduces their form drag by decreasing the curvature of surfaces and avoiding sudden changes of cross-sectional area and shape. Apart from the streamlining of aerofoil sections, where we look for a finer thickness/chord other parts of the airframe may also be streamlined, by adding fairings. Figure 7.19 shows how streamlining helps to substantially reduce form drag. The photographs shown in Figure 7.20 show how airflow can be studied with the use of a smoke generator; thin lines of smoke are drawn through a wind tunnel and pass over the object being investigated. This illustrates how a streamlined shape maintains the laminar flow while the cylinder produces eddies and a greater turbulent wake.

Interference drag

The total drag acting on an aircraft is greater than the sum of the component drag. This is because, due to flow, interference occurs at the various junctions of the surfaces. These include the wing/fuselage junctions, wing/engine-pylon junctions and those between tail plane, fin and fuselage. This flow interference results in additional drag, which we call interference drag. As this type of drag is not directly associated with lift, it is another form of parasite drag.

7.19 Streamlining and form drag

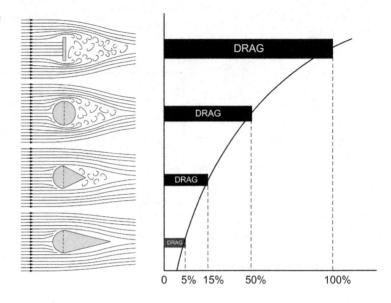

When airflow from the various aircraft surfaces meet, they form a wake behind the aircraft. The additional turbulence that occurs in the wake causes a greater pressure difference between the front and rear surfaces of the aircraft and therefore increases drag. As mentioned earlier, interference drag can be minimized by using suitable fillets, fairings and streamlined shapes.

Induced drag

Induced drag results from the production of lift. It is created by differential pressures acting on the top and bottom surfaces of the wing. The pressure above the wing is slightly below atmospheric, while the pressure beneath the wing is at or slightly above atmospheric. This results in the migration of the airflow at the wing tips from the high-pressure side to the low-pressure side. Since this flow of air is span-wise, it results in an overflow at the wing tip that sets up a spiralling action causing induced drag – see Figure 7.21. This whirling of the air at the wing tip is known as a vortex. Also, the air on the upper surface of the wing tends to move towards the fuselage and off the trailing edge. This air current forms a similar vortex at the inner portion of the trailing edge of the wing. These vortices increase drag due to the turbulence produced and this type of drag is known as induced drag.

7.20 Smoke generator: flows over a cylinder/aerofoil

Lower pressure

Higher pressure

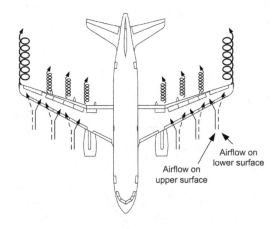

Airflow on
lower surface

Airflow on
upper surface

7.21 Wingtip vortices and induced drag

In the same way as lift increases with increase in AOA, so too does induced drag. This results from the greater pressure difference produced with increased AOA, creating even more violent vortices, greater turbulence and greater downwash of air behind the trailing edge. These vortices are visible on cool moist days when condensation takes place in the twisting vortices and they can be seen from the ground as vortex spirals.

If the speed of the aircraft is increased then lift will be increased. Thus to maintain straight-and-level flight the AOA of the aircraft must also be reduced. We have seen that increase in AOA increases induced drag. Therefore, by reducing the AOA and increasing speed, we reduce the ferocity of the wing-tip vortices and so reduce the induced drag. This is the direct opposite to form drag, which clearly increases with increase in velocity. In fact it can be shown that induced drag reduces in proportion to the square of the airspeed, while profile drag increases proportionally with the square of the airspeed.

Wing-tip stall

A situation can occur when an aircraft is flying at high AOA, e.g. on the approach to landing, where due to losses incurred by strong wing-tip vortices, one wing tip may stall while the remainder of the main plane is still lifting. This will result in more lift being produced by one wing than the other, resulting in a roll motion towards the stalled wing tip. This is most undesirable under any conditions and methods have been adopted to reduce losses at the wing tip. Three of the most common methods for reducing induced drag and wing-tip stall are to use one or more of the following design approaches for the wing:

* Washout
* Leading-edge spoilers
* Long narrow tapered wings

If the angle of incidence (AOI) of the wing is decreased towards the wing tip, there will be fewer tendencies for wing-tip vortices to form at high AOA, due to the fact that the wing tip is at a lower AOA than the remaining part of the wing. This design method is known as wash-out and the opposite i.e., an increase in the AOI towards the wing tip, is known as wash-in. Some aircraft are fitted with fixed spoilers to their inboard leading edge. These have the effect of disturbing the airflow and inducing the stall over the inboard section of the wing, before it occurs at the wing tip, thus removing the possibility of sudden wing-tip stall.

Another method of reducing induced drag is to have long narrow tapered wings, i.e. wings with a high aspect ratio. Unfortunately, from a structural point of view a long narrow tapered wing is quite difficult to build and this is often the limiting factor in developing high-aspect-ratio wings – Figure 7.22. The result of this design is to create smaller vortices that are a long way apart and therefore will not readily interact. The tapered end of a long thin wing, such as those fitted to a glider, helps reduce the strength of the wing-tip vortices and so induced drag. Aspect ratio may be calculated using any one of the following three expressions, dependent on the information available:

* Span/mean chord
* $Span^2$/area
* Wing area/mean $chord^2$

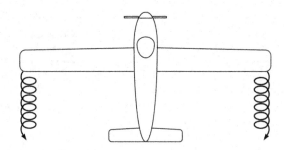

7.22 High-aspect-ratio wings and induced drag

Total drag

Knowing the circumstances under which the total drag of an aircraft is at a minimum reduces fuel burn, and improves aircraft performance and operating costs. We know that profile drag increases with the square of the airspeed and that induced drag decreases with the square of the airspeed. Therefore there must be an occasion when at a particular airspeed and AOA, drag is at a minimum. The drag curves for induced drag and profile drag – Figure 7.23 – show when their combination, i.e. total drag, is at a minimum.

FORCES ACTING ON AN AEROPLANE

We have already dealt with lift and drag when we considered aerofoil sections. We now look at these two forces and two others, thrust and weight, in particular with respect to their effect on the aircraft as a whole. For the aeroplane to maintain constant height the lift force created by the aerofoil sections must be balanced by the weight of the aircraft. Similarly, for an aeroplane to fly with constant velocity, or zero acceleration, the thrust force must be equal to the drag force that opposes it.

Figure 7.24 shows the four flight forces acting at right angles to one another with their appropriate lines of action: lift, weight, thrust and drag. Lift of the main planes acts perpendicular to the relative airflow through the CP of the main aerofoil sections. Weight acts vertically downwards through the aircraft's CG. Thrust of the engines works along the engine axis approximately parallel to the direction of flight. Drag is the component acting rearwards parallel to the direction of the relative airflow and is the resultant of

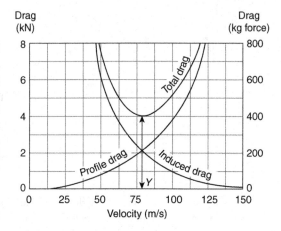

7.23 Induced drag and profile drag

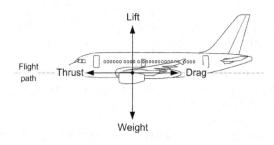

7.24 Four flight forces

two components: induced drag and profile drag. For convenience, the total drag is said to act at a point known as the centre of drag.

The weight which acts through the CG depends on every individual part of the aircraft and will vary depending on the distribution of passengers, crew, freight and fuel consumption. The line of action of the thrust is set in the basic design and is totally dependent on the position of the propeller shaft or the centre line of the exhaust jet.

The drag may be found by calculating its component parts separately or by experimenting with models in a wind tunnel. The four forces do not, therefore, necessarily act at the same point, so that equilibrium can only be maintained providing that the moments produced by the forces are in balance. In practice, the lift and weight forces may be so designed

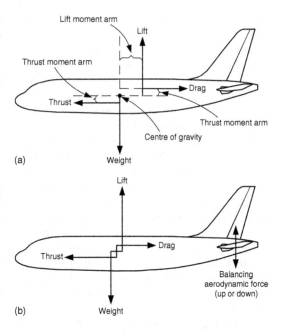

(a)

(b)

7.25 Force couples for straight-and-level flight

as to provide a nose-down couple (Figure 7.25), so that in the event of engine failure a nose-down gliding attitude is produced. For straight-and-level flight, the thrust and drag must provide an equal and opposite nose-up couple.

The design of an aircraft, however, will not always allow a high-drag and low-thrust line, so some other method of balancing the flight forces must be found.

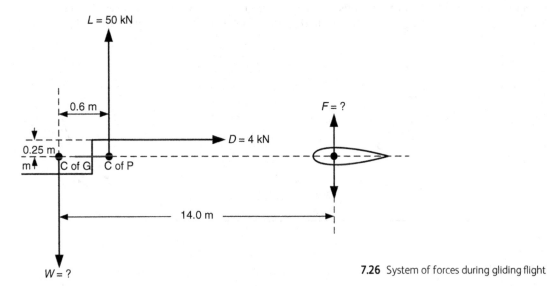

7.26 System of forces during gliding flight

This involves the use of the tail plane or horizontal stabilizer. One reason for fitting a tail plane is to counter the out-of-balance pitching moments that arise as a result of inequalities with the two main couples. The tail plane is altogether smaller than the wings – however, because it is positioned some distance behind the CG, it can exert considerable leverage from the moment produced (see Figure 7.26).

At high speed the AOA of the main plane will be small. This causes the CP to move rearwards, creating a nose-down pitching moment. To counteract this, the tail plane will have a downward force acting on it to rebalance the aircraft. Following the same argument for high AOA at slow speeds, the CP moves forward, creating a nose-up pitching moment. Thus, tail planes may need to be designed to carry loads in either direction. A suitable design for this purpose is the symmetrical cambered tail plane, which at zero AOA will allow the chord line of the section to be the neutral line.

Most tail planes have been designed to act at a specified AOA for normal flight modes. However, due to variables (such as speed), changing AOA with changing load distribution and other external factors, there are times when the tail plane will need to act with a different AOI – to allow for this some tail planes are moveable in flight and are known as the all-moving tail plane.

Flight forces in steady manoeuvres

We now consider the forces that act on the aircraft when gliding, diving, climbing and moving in a horizontal banked turn. Aircraft with zero thrust cannot maintain height indefinitely. Gliders or aircraft with total engine failure usually descend in a shallow flight path at a steady speed. The forces that act on an aircraft during gliding flight are shown in Figure 7.27. If the aircraft is descending at steady speed, we may assume that it is in equilibrium and a vector force triangle may be drawn as shown, where:

D = drag, W = weight, L = lift and γ = the glide angle.

Then from the vector triangle:

$Sin\ y = drag/weight$ and $cos\ y = lift/weight$

If an aircraft suffers a total loss of power and has less thrust than drag, it can only maintain constant speed by pitching nose down and adopting an optimum gliding angle.

Climbing flight

In a constant-speed climb, the thrust produced by the engines must be greater than the drag to maintain a steady speed. The steady-speed climb is illustrated in Figure 7.28(a), where again the vector triangle of the forces is given in Figure 7.28(b). From the vector triangle of forces:

$Sin\ y = (T\text{-}D)/W$ and $cos\ y = L/W$

If an aircraft is in a vertical climb at constant speed, the aircraft must have more thrust than weight (W) in

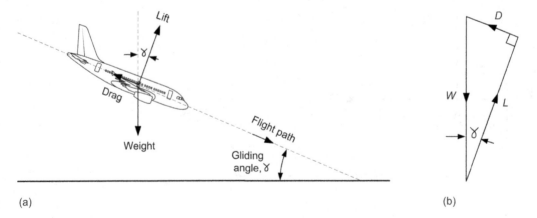

(a)

(b)

7.27 Gliding flight

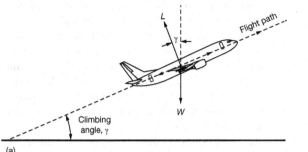

(a)

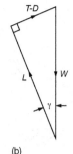

7.28 Forces acting on an aircraft during steady-speed climb

(b)

order to overcome the drag (D), i.e.: Thrust = W + D (for steady vertical climb where the lift is zero).

Turning flight

When gliding, diving and climbing, the aircraft has been in equilibrium, where its speed and direction were fixed. If the aircraft manoeuvres by changing speed or direction, acceleration takes place and equilibrium is lost. When an aircraft turns, centripetal force is required to act towards the centre of the turn, in order to hold the aircraft in the turn (Figure 7.29). This centripetal force must be balanced by the lift component in order to maintain a constant radius (steady) turn; this is achieved by banking the aircraft. In a correctly banked turn, the forces are as shown in Figure 7.30. The horizontal component of lift is equal to the centrifugal force, holding the aircraft in the turn.

FLIGHT STABILITY AND DYNAMICS

Stability

The stability of an aircraft is a measure of its tendency to return to its original flight path after a displacement. This displacement caused by a disturbance can take place in any of three axes of reference: longitudinal, normal and lateral (see Figure 7.31). These axes are imaginary lines passing through the CG of the aircraft, which are mutually perpendicular to one another, i.e. they are at right angles to one another. All the complex dynamics concerned with aircraft use these axes to model and mathematically define stability and control parameters.

Any object which is in equilibrium when displaced by a disturbing force will react in one of three ways once the disturbing force is removed, thus:

- When the force is removed and the object returns to the equilibrium position, it is said to be stable.

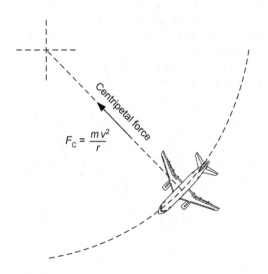

7.29 Centripetal force during a turn

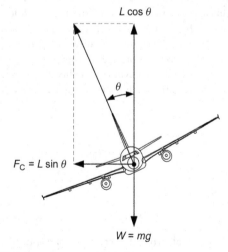

7.30 Forces during a correctly banked turn

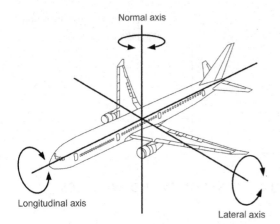

7.31 Aircraft axes and planes of reference

- When the force is removed and the object continues to move in the direction of the force and never returns to the equilibrium position, it is said to be unstable.
- When the force is removed and the object stops in the position to which it has been moved to, neither returning nor continuing, it is said to be neutrally stable.

These reactions are illustrated in Figure 7.32, where a ball bearing is displaced and then released. In Figure 7.32(a), it can be clearly seen that the ball bearing, once released in the bowl, will gradually settle back to the equilibrium point after disturbance. In Figure 7.32(b), it can be seen that, due to the effects of gravity, the ball bearing will never return to its original equilibrium position on top of the cone. While in Figure 7.32(c), after disturbance, the ball bearing eventually settles back into a neutral position somewhere remote from its original resting place.

There are in fact two types of stability that we need to consider – static stability and dynamic stability. An object such as an aircraft is said to have static stability if, once the disturbing force ceases, it starts to return to the equilibrium position. With respect to dynamic stability, consider again the situation with the ball bearing in the bowl, where it is statically stable and starts to return to the equilibrium position. In returning, the ball bearing oscillates backwards and forwards before it settles. This oscillation is damped out and grows smaller until the ball bearing finally returns to the equilibrium position. An object is said to be dynamically stable if it returns to the equilibrium position, after a disturbance, with decreasing oscillations. If an increasing oscillation occurs then the object may be statically stable but dynamically unstable. This is a very dangerous situation which can happen to moving objects, if the force balance is incorrect. An example of dynamic instability is helicopter rotor vibration, if the blades are not properly balanced.

Aircraft-stability dynamics

The static and dynamic responses of an aircraft, after it has been disturbed by a small force, are represented by the series of diagrams shown in Figure 7.33. In the first example, Figure 7.33(a), this illustrates deadbeat static stability, where the aircraft returns to the equilibrium position, without any dynamic oscillation, caused by the velocities of motion. Figure 7.33(b) shows the situation for an aircraft that is both statically and dynamically stable (equivalent to the ball bearing in the bowl). Under these circumstances, the aircraft will return to its equilibrium position after a few decreasing oscillations. Figure 7.33(c) illustrates the undesirable situation where the aircraft may be statically stable but is dynamically unstable – in other words, the aircraft is out of control.

> ### KEY POINT
>
> It is not possible for an aircraft to be statically unstable and dynamically stable, but the reverse situation is possible.

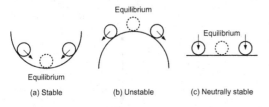

7.32 Reaction of an object after removal of a disturbing forces: (a) stable (b) unstable (c) neutral

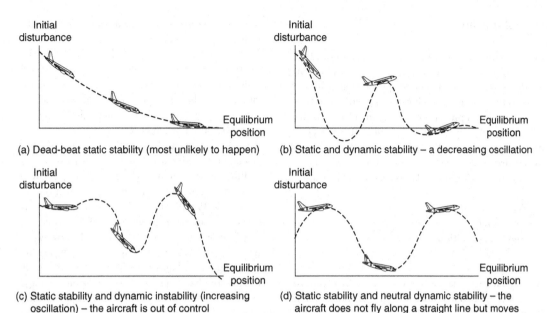

(a) Dead-beat static stability (most unlikely to happen)

(b) Static and dynamic stability – a decreasing oscillation

(c) Static stability and dynamic instability (increasing oscillation) – the aircraft is out of control

(d) Static stability and neutral dynamic stability – the aircraft does not fly along a straight line but moves slowly up and down. A *phugoid* oscillation.

7.33 Static and dynamic responses of an aircraft

Figure 7.33(d) illustrates the situation for an aircraft that has static stability and neutral dynamic stability. Under these circumstances the aircraft does not fly in a straight line but is subject to very large, but low-frequency, reactions known as phugoid oscillations.

When considering stability, we assume that the CG of the aircraft continues to move in a straight line and that the disturbances to be overcome cause rotational movements about the CG. These movements can be:

- Rolling movements about (around) the longitudinal axis: lateral stability;
- Yawing movements about the normal axis: directional stability;
- Pitching movements about the lateral axis: longitudinal stability.

Figure 7.34 illustrates three directions of movement that must be damped, if the aircraft is to be considered stable: roll, yaw and pitch. The lateral stability is the inherent ability of an aircraft to recover from a disturbance around the longitudinal plane (axis), i.e. rolling movements. Similarly, longitudinal stability is the inherent (built-in) ability of the aircraft to recover from disturbances around the lateral axis i.e. pitching movements. Finally, directional stability is the inher-

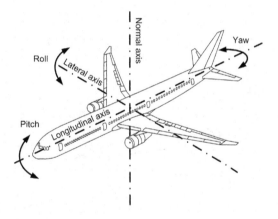

7.34 Roll, yaw and pitch

ent ability of the aircraft to recover from disturbances around the normal axis.

There are many aircraft features specifically designed to either aid stability or reduce the amount of inherent stability an aircraft possesses, depending on aircraft configuration and function. We look at some of these design features next when we consider lateral, longitudinal and directional stability in a little more detail.

Lateral stability

From what has been said above, an aircraft has lateral stability, following a roll displacement, a restoring moment is produced which opposes the roll and returns the aircraft to a wings-level position. In that, aerodynamic coupling produces rolling moments that can set up side-slip or yawing motion. It is therefore necessary to consider these interactions when designing an aircraft to be inherently statically stable in roll. The main contributors to lateral static stability are:

- Wing dihedral
- Sweepback
- High wing position
- Keel surface.

A design feature that has the opposite effect to those given above, i.e. that reduces stability, is anhedral. The need to reduce lateral stability may seem strange, but combat aircraft and many high-speed automatically controlled aircraft use anhedral design features to provide more manoeuvrability.

Dihedral angle is the upward inclination of the wings from the horizontal. The amount of dihedral angle is dependent on aircraft type and wing configuration, i.e. whether the wings are positioned high or low with respect to the fuselage and whether or not they are straight or swept back.

The righting effect from a roll using wing dihedral angle may be considered as a two-stage process; where the rolling motion is first stopped and then the down-going wing is returned to the horizontal position.

Figure 7.35 illustrates how roll stability is achieved; we can see that for an aircraft in a roll, one wing will move down and the other will move up – Figure 7.35(a) – as a result of the rolling motion. The vector diagrams – Figure 7.35(b) – show the AOA resultants for the up-going and down-going wings. The direction of the free-stream airflow approaching the wing is changed and the AOA on the down-going wing is increased, while the AOA on the up-going

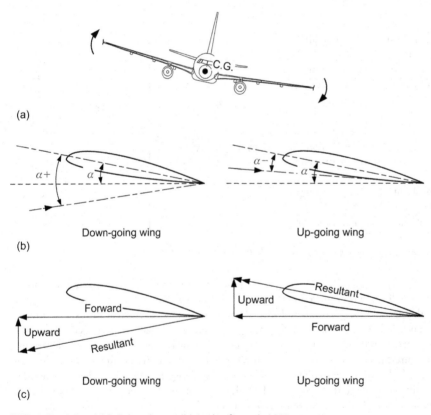

(a)

(b)

Down-going wing Up-going wing

(c)

Down-going wing Up-going wing

7.35 Roll stability: (a) Roll disturbance; (b) Angle of attack; (c) Vectors

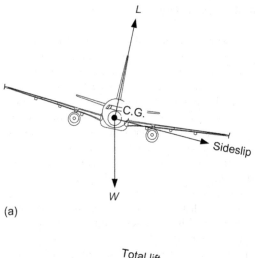

(a)

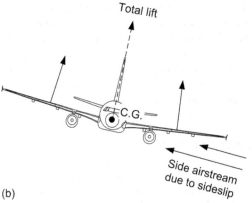

(b)

7.36 Achieving lateral stability using wing dihedral

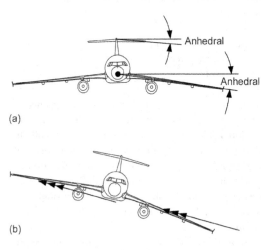

(a)

(b)

7.37 Anhedral and lateral stability

wing is decreased. Referring to Figure 7.35(c), this causes a larger C_L and lift force to be produced on the lower wing and a smaller lift force on the upper wing, so the roll is stopped. When the roll stops the lift forces equalize again and the restoring effect is lost.

Inherent lateral stability, i.e. naturally returning the aircraft to the equilibrium position, can be achieved with wing dihedral (Figure 7.36). A natural consequence of banking the aircraft is to produce a component of lift which acts in such a way as to cause the aircraft to sideslip. In Figure 7.36(a) the component of lift resulting from the angle of bank can clearly be seen. It is this force that is responsible for side-slip. Now if the wings were straight the aircraft would continue to side-slip, but if dihedral angle is built in, the sideways airstream will create a greater lift force on the down-going wing, as seen in Figure 7.36(b). This difference in lift force will restore the aircraft until it is no longer banked over and side-slipping stops.

Anhedral is the downward inclination of the wings – Figure 7.37(a) – and is used to decrease lateral stability. In this case, as the aircraft side-slips, the lower wing, due to its anhedral, will meet the relative airflow at a reduced AOA – Figure 7.37(b) – so reducing lift, while the upper wing will meet the relative airflow at a higher AOA and will produce even more lift. The net effect will be to increase the roll and thus reduce lateral stability.

High wing and keel surface

Aircraft designed with a high wing for roll stability (Figure 7.38) have the centre of lift above the fuselage, and the CG in a low position within the fuselage; this creates a pendulum effect in a side-slip. The wing and body drag, resulting from the relative airflow in the side-slip and the forward motion of the aircraft, produces forces that act parallel to the longitudinal axis and at right angles to it in the direction of the raised wing. These forces produce a turning moment about the CG, which together with a certain loss of lift on the upper main plane (caused by turbulence over the fuselage) and the pendulum effect tends to lift the aircraft and correct the roll disturbance.

Again, when an aircraft is in a side-slip as a result of a roll, air loads will act on the side of the fuselage and on the vertical stabilizer (fin/rudder assembly) which together form the keel surface, i.e.

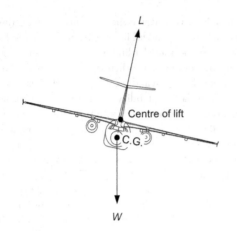

7.38 High wing for roll stability

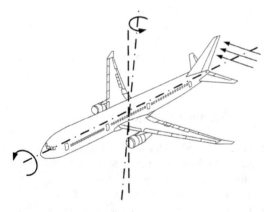

7.39 Relative airflow acting on the fin

the cross-sectional area of the aircraft when viewed from the side. These loads produce a rolling moment, which (in addition to the aircraft's lateral stability) is further corrected by a restoring moment created by relative airflow acting on the fin. The magnitude of this moment is dependent on the size of the fin and its distance from the aircraft CG – see Figure 7.39.

Sweepback and lateral stability

Wings with sweepback can also enhance lateral stability – Figure 7.40(a). As the aircraft sideslips following a disturbance in roll, the lower sweptback wing generates more lift than the upper wing. This results from the fact that in the sideslip the lower wing presents more of its span to the airflow than the upper wing; therefore the lower wing generates more lift and tends to restore the aircraft to a wings-level position.

Figure 7.40(b) shows how the component of the velocity perpendicular to the leading edge is increased on the down-going main-plane. It is this component of velocity that produces the increased lift and together with the increase in effective wing span restores the aircraft to a wings-level position.

In addition, the surface of the down-going main-plane will be more steeply cambered to the relative airflow than that of the up-going main-plane. This will result in the down-going main-plane having a higher lift coefficient compared with the up-going main-plane during side-slip, thus the aircraft tends to be restored to its original attitude.

Lateral dynamic stability

The relative effect of combined rolling, yawing and side-slip motions, resulting from aerodynamic coupling (see directional stability), determine the lateral dynamic stability of an aircraft. If the aircraft-stability characteristics are not sufficient, the complex motion interactions produce three possible types of instability – these are:

- Directional divergence
- Spiral divergence
- Dutch roll

If an aircraft is directionally unstable a divergence in yaw may result from an unwanted yaw disturbance. In addition a side force will act on the aircraft while in the yawed position; it will curve away from its original flight path. If under these circumstances the aircraft has lateral static stability, directional divergence will occur without any significant degree of bank angle and the aircraft will still fly in a curved path with a large amount of side-slip.

Spiral divergence exists when directional static stability is very large compared to lateral stability.

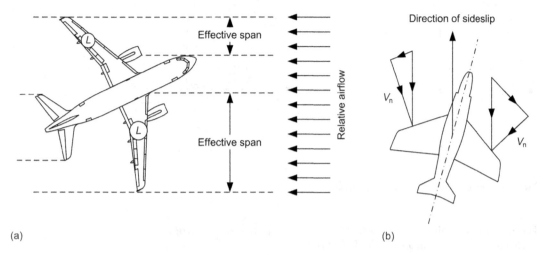

(a)   (b)

7.40 Sweepback and lateral stability

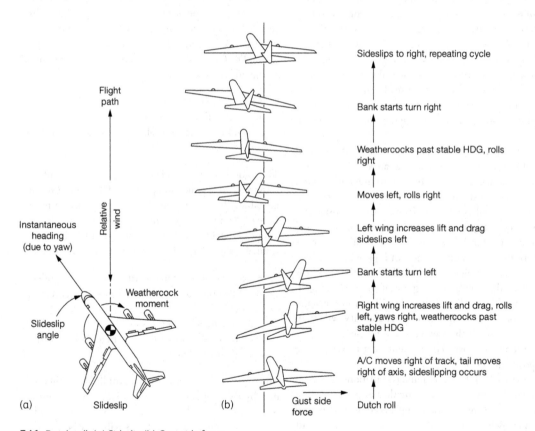

(a)   Slideslip   (b)   Gust side force   Dutch roll

7.41 Dutch roll: (a) Sideslip (b) Gust side force

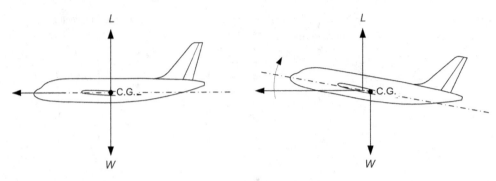

7.42 Disturbance causing the nose to pitch up

This may occur on aircraft with a significant amount of anhedral coupled with a large fin. If an aircraft is subject to a yaw displacement, then because of the greater directional stability the yaw would be quickly eliminated by the stabilizing yawing moment set up by the fin. However, a rolling moment would also be set up in the same direction as the yaw and if this rolling moment were strong enough to overcome the restoring moment due to static stability, the angle of bank would increase and cause the aircraft nose to drop into the direction of the yaw. The aircraft then begins a nose spiral which may develop into a spiral dive.

Dutch roll (Figure 7.41) is an oscillatory mode of instability which may occur if an aircraft has positive directional static stability, but not so much in relation to static lateral stability, as to lead to spiral divergence. Thus Dutch roll is a form of lateral dynamic instability that does not quite have the inherent dangers associated with spiral divergence. Dutch roll may occur where there is a combination of high wing loading, sweepback and high altitude and where the weight is distributed towards the wing tips. If an aircraft is again subject to a yaw disturbance, it will roll in the same direction as the yaw. Directional stability will then begin to reduce the yaw and due to inertia forces, the aircraft will over-correct and start to yaw in the opposite direction.

Now each of these continuing oscillations in yaw act in such a manner as to cause further displacements in roll, the resulting motion being a combination of roll and yaw oscillations which have the same frequency but are out of phase with each other. The development of Dutch roll is prevented by fitting aircraft with yaw-damping systems, which will be looked at in more detail in the chapter covering autopilots.

Longitudinal stability

As previously described, an aircraft is longitudinally statically stable if it has the tendency to return to a trimmed AOA position following a pitching disturbance. Consider an aircraft that is subject to a disturbance causing the nose to pitch up (Figure 7.42). The CG will continue to move in a straight line, so the effects will be:

- An increase in the AOA
- The CP will move forward
- A clockwise moment about the CG provided by the lift force

This causes the nose to keep rising so that it will not return to the equilibrium position. The aircraft is thus unstable. (If the pitching disturbance causes a nose-down attitude, the CP moves to the rear and the aircraft is again unstable.) For an aircraft to be longitudinally statically stable, it must meet two criteria:

- A nose-down pitching disturbance must produce aerodynamic forces to give a nose-up restoring moment.
- This restoring moment must be large enough to return the aircraft to the trimmed AOA position after the disturbance.

Thus, the requirements for longitudinal stability are met by the tailplane (horizontal stabilizer). The CG of the aircraft will still continue to move around a vertical straight line. The effects will now be:

- An increase in AOA for both wing and tailplane;
- The CP will move forward and a lift force will be produced by the tailplane;

- The tailplane will provide an anticlockwise restoring moment, i.e. greater than the clockwise moment, produced by the wing-lift force as the CP moves forward.

A similar restoring moment is produced for a nose-down disturbance except that the tailplane lift force acts downwards and the direction of the moments are reversed. From the above argument, it can be seen that the restoring moment depends on:

- The size of the tailplane (or horizontal stabilizer)
- The distance of the tailplane behind the CG
- The amount of elevator movement (or complete tailplane movement, in the case of aircraft with all-moving slab tailplanes) which can be used to increase tailplane lift force.

All of the above factors are limited and there will be a limit to the restoring moment that can be applied. It is therefore necessary to ensure that the disturbing moment produced by the wing-lift moment about the CG is also limited. This moment is affected by movements of the CG due to differing loads and load distributions, in addition to fuel-load distribution.

It is therefore vitally important that the aircraft is always loaded within the CG limits specified in the aircraft weight and balance documentation. The result of failure to observe these limits may result in the aircraft becoming unstable, with subsequent loss of control.

Dynamic longitudinal stability

The first mode of dynamic longitudinal stability is phugoid motion; this consists of long-period oscillations that involve noticeable changes in pitch attitude, aircraft altitude and airspeed. The pitching rate is low and because only very small changes in AOA occur, damping is weak and sometimes negative.

The second mode of dynamic longitudinal stability involves a short period of motion of relatively high frequency that involves negligible changes in aircraft velocity. During this type of motion, static longitudinal stability restores the aircraft to equilibrium and the amplitude of the oscillation is reduced by the pitch damping contributed by the tailplane (horizontal stabilizer). If instability was to exist in this mode of oscillation, porpoising of the aircraft would occur and

because of the relative high frequency of oscillation, the amplitude could reach dangerously high proportions, with severe flight loads being imposed on the structure.

Directional stability

As already discussed, directional stability of an aircraft is its inherent (built-in) ability to recover from a disturbance in the yawing plane, i.e. about the normal axis. However, unlike longitudinal stability, it is not independent in its influence on aircraft behaviour because, as a result of what is known as aerodynamic coupling, yaw-displacement moments also produce roll-displacement moments about the longitudinal axis. As a consequence of this aerodynamic coupling, aircraft directional motions have an effect on lateral motions and vice versa. The nature of these motions is yawing, rolling and side-slip, or any combination of the three.

With respect to yawing motion only, the primary influence on directional stability is provided by the fin (or vertical stabilizer). As the aircraft is disturbed from its straight-and-level path by the nose or tail being pushed sideways (yawed), due to its inertia the aircraft will continue to move in the direction created by the disturbance. This will expose the keel surface to the oncoming airflow. The fin, acting as a vertical aerofoil, will generate a sideways lift force which tends to swing the fin back towards its original position, straightening the nose as it does so.

The restoring moment created by the fin after a yawing disturbance is shown in Figure 7.43. The greater the keel surface area (which includes the area

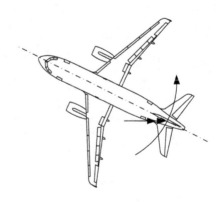

7.43 Restoring moment created by the fin

of the fin) behind the CG, and the greater the moment arm, then the greater will be the directional stability of the aircraft. Knowing this, it can be seen that a forward CG is preferable to an aft CG, since it provides a longer moment arm for the fin.

CONTROL AND CONTROLLABILITY

Introduction

In addition to varying degrees of stability, an aircraft must have the ability to respond to the requirements for manoeuvring and trimming about its three axes. Thus the aircraft must have the capacity for the pilot to control it in roll, pitch and yaw, so that all desired flying attitudes may be achieved throughout all phases of flight.

Controllability is a different consideration to stability in that it requires aerodynamic forces and moments to be produced about the three axes, where these forces always oppose the inherent stability-restoring moments, when causing the aircraft to deviate from its equilibrium position. Thus if an aircraft is highly stable, the forces required to deviate the aircraft from its current position will need to be greater than those required to act against an aircraft that is less inherently stable. This is one of the reasons why aircraft that are required to be highly manoeu-vrable and respond quickly to pilot or autopilot demands are often designed with an element of instability built in.

Different control surfaces are used to control the aircraft about each of the three axes. Movement of these control surfaces changes the airflow over the

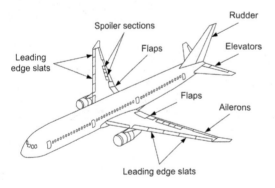

7.44 Conventional flying control surfaces

aircraft's surface. This, in turn, produces changes in the balance of the forces that keep the aircraft flying straight and level, thus creating the desired change necessary to manoeuvre the aircraft. Conventional flying-control surfaces are always designed so that each gives control about the aircraft axes as shown in Figure 7.44; thus:

* Ailerons provide roll control about the longitu-dinal axis.
* Elevators provide longitudinal control in pitch about the lateral axis.
* The rudder gives yaw control about the normal axis.

Another common control grouping is the taileron, which provides the combined control functions of the tailplane and ailerons. Here the two sides of the slab tail will act collectively to provide tailplane control in pitch or differentially to provide aileron control.

Ailerons

Aileron movements cause the aircraft to roll by producing a difference in the lift forces over the two wings; one aileron moves up and the other simulta-neously moves down. The aileron that is controlled upwards causes an effective decrease in the angle of attack (Figure 7.45) of the wing, with a subsequent reduction in C_L and lift force, thereby causing a down-going wing. Similarly an aileron deflected downwards causes an effective increase in the AOA of the wing, increasing C_L and lift force, so we have an up-going wing – Figure 7.46. While the ailerons remain deflected the aircraft will continue to roll. To maintain a steady angle of bank, the ailerons must be returned to the neutral position after the required angle has been reached.

Ailerons not only produce a difference in lift forces between the wings, but also a difference in drag force. The drag force on the up-going aileron (down-going wing) becomes greater due to air loads and turbulence than the drag force on the down-going wing. The effect of this aileron drag is to produce an adverse yaw away from the direction of turn (Figure 7.47).

There are two common methods of reducing aileron drag. The first involves the use of differential ailerons (Figure 7.48), where the aileron that is deflected downwards moves through a smaller angle

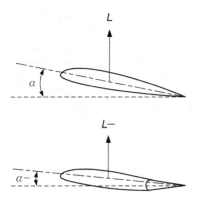

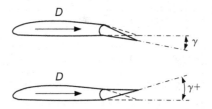

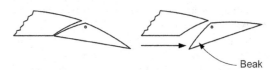

7.48 Differential ailerons/reduced aileron drag

7.49 Frise ailerons

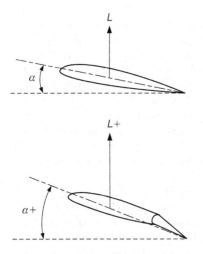

7.45 Up-going aileron/down-going wing

7.46 Down-going aileron/up-going wing

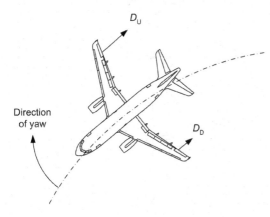

7.47 Aileron drag and adverse yaw

than the aileron that is deflected upwards. This tends to equalize the drag on the two wings.

The second method, normally found on older low-speed aircraft, uses Frise ailerons (Figure 7.49). A Frise aileron, named after engineer Leslie George Frise (1897–1979), is pivoted as shown to produce a 'beak', which projects downwards into the airflow, when the aileron is deflected upwards, but does not project into the airflow when it is deflected downwards. The beak causes an increase in drag on the down-going wing, helping to equalize the drag between the wings.

Aileron reversal

At low speeds, an aircraft has a relatively high AOA that is close to the stall angle. If the ailerons are operated while the wings are at this high angle of attack, the increase in the effective angle of attack may cause the wing with the aileron deflected downwards to have a lower C_L than the other, instead of the normal higher C_L. This will cause the wing to drop instead of rise and the aircraft is said to have suffered low-speed aileron reversal.

When ailerons are deflected at high speeds the aerodynamic forces set up may be sufficiently large to twist the outer end of the wing (Figure 7.50). This can cause the position of the chord line to alter so that the result is the opposite of what would be expected. That is, a downward deflection of the aileron causes the wing to drop and an upward deflection causes the

7.50 Ailerons deflected at high speeds

wing to rise – under these circumstances we say that the aircraft has suffered a high-speed aileron reversal. This is a problem for high-speed military and large transport aircraft; solutions include:

• Wings featuring stiff structural construction
• Two sets of ailerons
• Spoilers

Designing 'stiff' wings resists torsional divergence beyond the maximum speed of the aircraft. Two sets of ailerons, one outboard pair that operate at low speeds and one inboard pair that operate at high speeds, means that the twisting moment will be less than when the ailerons are positioned outboard. The use of spoilers, operating either independently or in conjunction with ailerons, reduces the lift on the down-going wing by interrupting the airflow over the top surface. Spoilers do not cause the same torsional divergence of the wing and have the additional advantage of providing increased drag on the down-going wing, thus helping the adverse yaw problem created by aileron drag.

Rudder

Movement of the rudder creates a lift force which yaws the aircraft nose. Although this will cause the aircraft to turn, eventually, it is much more effective to use the ailerons to bank the aircraft, with minimal use of the rudder. The rudder is typically used during take-off and landing to keep the aircraft straight while on the runway. (With the aircraft on the ground the ailerons have no effect.) During the turn, the rudder provides control assistance to carry out coordinated turns. The rudder has applications at low speeds and high angles of attack to help raise a dropping wing that has suffered aileron reversal. On multi-engine aircraft, the rudder is used to correct yawing when asymmetric power conditions exist.

Elevators

When the elevator is controlled upwards it causes a downward lift force on the tailplane (Figure 7.51). This upward movement of the elevators causes an increase in the AOA and the forward movement of the CP. This means that the aircraft will pitch upwards if the speed is maintained or increased because the C_L rises as the AOA rises and a larger lift force is created. To maintain an aircraft in a steady climb the elevators are returned to the neutral position. If elevator deflection is maintained, the aircraft will continue up into a loop.

If the speed is reduced as the elevators are raised, the aircraft continues to fly level because the increase in C_L due to the increased AOA is balanced by the decrease in velocity – so from the formula for lift, total lift force will remain the same. Under these circumstances, the elevator deflection must be maintained to keep the nose high unless the speed is

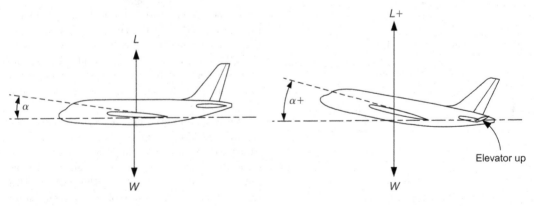

7.51 Elevator control

increased again. If we wish to pull an aircraft out of a glide or dive, then upward elevator movement must be used to increase the total lift force, necessary for this manoeuvre.

LIFT AUGMENTATION DEVICES

Introduction

When aircraft takes off and/or lands in a relatively short distance, its wings must produce sufficient lift at a much slower speed than in normal cruising flight. (This can also apply to aircraft that fly low speeds as part of their role, such as search and rescue.) It is also necessary during landing to have some means of slowing the aircraft down. Both these requirements can be met by the use of flaps and slats or a combination of both.

Flaps are essentially moving wing sections which increase wing camber and therefore angle of attack. In addition, in some cases, the effective wing area is also increased. Dependent on type and complexity, flap systems are capable of increasing C_{Lmax} by up to approximately 90 per cent of the clean wing value.

Flaps also greatly increase the drag on the wings, thus slowing the aircraft down. Thus on take-off, flaps are partially deployed and the increase in drag is overcome with more thrust, while on landing they are fully deployed for maximum effect.

Trailing edge flaps

There are many types of trailing-edge flaps; the more common types are described below.

The plain flap – Figure 7.52 – is normally retracted to form a complete section of trailing edge, and hinged downward in use. The split flap – Figure 7.53 – is formed by the hinged lower part of the trailing edge only. When the flap is lowered the top surface is

unchanged, thus eliminating airflow breakaway which occurs over the top surface of the plain flap at large angles of depression. The slotted flap – Figure 7.54 – has a gap or slot formed between the wing and flap. Air flows through the gap from the lower surface and over the top surface of the flap. This increases lift by speeding up the airflow over the top surface of the flap. This more energetic laminar flow remains in contact with the top surface of the flap for longer, delaying boundary-layer separation and maintaining a high degree of lift.

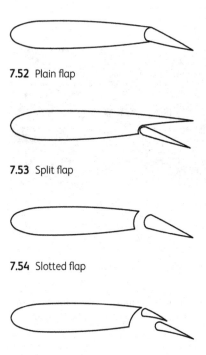

7.52 Plain flap

7.53 Split flap

7.54 Slotted flap

7.55 Fowler flap

The Fowler flap, named after Harlan Fowler (1895–1982), an aeronautical engineer and inventor, is similar to the split flap, but this type of flap – see Figure 7.55 – moves rearwards as well as downwards on tracks, creating slots and thereby increasing both wing camber and wing area. More than one flap of this type can be connected as part of this design. In the blown flap (Figure 7.56), air bled from the engines is ducted over the top surface of the flap to mix with and re-energize the existing airflow.

Figure 7.57 shows the trailing-edge flap system of a Boeing 747-400, in the deployed position. This system is a multi-slotted Fowler combination, which

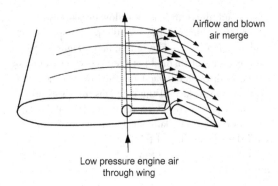

Airflow and blown air merge

Low pressure engine air through wing

7.56 Blown flap

7.57 Trailing-edge flap system of a Boeing 747-400

combines and enhances the individual attributes of the slotted flap and Fowler flap, greatly enhancing lift.

Leading-edge flaps

As mentioned earlier, leading-edge flaps – Figure 7.58 – are used to augment low-speed lift, especially on swept-wing aircraft. This type of flap, developed by a German engineer called Werner Krueger (1910–2003), further increases the wing's camber and is normally coupled to operate together with trailing-

7.58 Kruger flap

edge flaps. The Krueger flap also prevents leading-edge separation that takes place on thin sharp-edged wings at high angles of attack.

Leading edge slats and slots

Leading-edge slats are small aerofoils (Figure 7.59) fitted to the wing-leading edges. When open, slats form a slot between themselves and the wing, through which air from the higher pressure lower surface accelerates and flows over the wing-top surface to maintain lift and increase the stalling angle of the wing. Slats may be fixed, controlled or automatic.

7.59 Leading-edge slat

A wing-tip slot is a suitably shaped aperture built into the wing structure near the leading edge – Figure 7.60. Slots guide and accelerate air from below the wing and discharge it over the upper surface to re-energize the existing airflow. Slots may be fixed, controlled, automatic or blown.

Some aircraft are designed with a combination of lift-augmentation devices. In Figure 7.61 a typical leading- and trailing-edge lift-enhancement system is illustrated. This system consists of a triple-slotted Fowler flap at the trailing edge, with a slat and Kruger flap at the leading edge. This combination will significantly increase the lift capability of the aircraft.

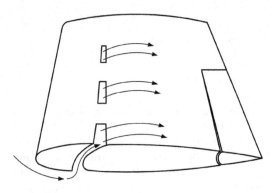

7.60 Wing-tip slot

7.61 Combination of lift-augmentation devices

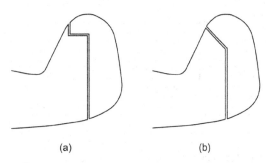

(a)   (b)

7.63 Horn balance: (a) Standard horn balance; (b) Graduated horn balance

BALANCE, MASS BALANCE AND CONTROL-SURFACE TABS

Aerodynamic balance

On all but the smallest of low-speed aircraft, the size of the control force will produce hinge moments that produce control-column forces that are too high for easy control operation. Non-sophisticated light aircraft do not necessarily have the advantage of powered controls and, as such, they are usually designed with some form of inherent aerodynamic balance that assists the pilot during their operation. There are several methods of providing aerodynamic balance; three such methods are given below.

Inset hinges are set back in the control surface so that the airflow strikes the surface in front of the hinge to assist with control movement – Figure 7.62. A rule of thumb with this particular design is to limit the amount of control surface forward of the hinge line to about 20 per cent of the overall control-surface area. Following this rule helps prevent excessive snatch and over-balance.

Another way of achieving lower hinge moments and so assisting the pilot to move the controls is to use a horn balance (Figure 7.63). The principle of operation is the same as for the insert hinge. Horn balances can be fitted to any of the primary flying-control surfaces. Figure 7.63(a) shows the standard horn balance – this device is sometimes prone to snatch. A graduated horn balance – Figure 7.63(b) – overcomes this problem by introducing a progressively increasing amount of control-surface area into the airflow forward of the hinge line, rather than a sudden change in area that may occur with the standard horn balance.

The internal balance principle is illustrated in Figure 7.64 – note the balancing area is inside the wing. Downward movement of the control surface creates a decrease in pressure above the wing and a relative increase below the wing. Since the gap between the low- and high-pressure areas is sealed (using a flexible strip), the pressure acting on the strip creates a force that acts upwards, which in turn produces the balancing moment that assists the pilot to move the controls further. This situation would obviously work in reverse when the control surface is moved up.

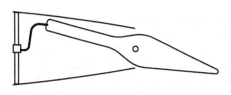

7.64 Internal balance principle

Mass balance

If the CG of a control surface is some distance behind the hinge line, then the inertia of the control surface may cause it to oscillate about the hinge, as the structure distorts during flight. This undesirable situation is referred to as control-surface flutter and in certain

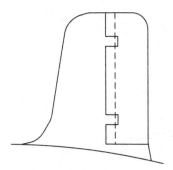

7.62 Inset hinges in the control surface

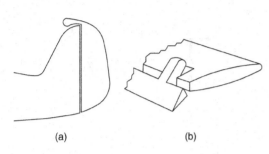

(a)   (b)

7.65 Mass balance

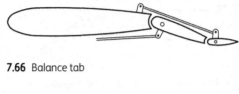

7.66 Balance tab

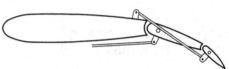

7.67 Anti-balance tab

circumstances these flutter oscillations can be so severe as to cause damage or possible failure of the structure. Flutter may be prevented by adding a carefully determined mass to the control surface in order to bring the CG closer to the hinge line. This procedure, known as mass balance, helps reduce the inertia moments and so prevents flutter developing. Figure 7.65 shows examples of mass-balance arrangements, where the mass is adjusted forward of the hinge line, as necessary. Control surfaces that have been resprayed or repaired must be check weighed and the CG recalculated to ensure that it remains within specified limits.

Tabs

A tab is a small hinged surface forming part of the trailing edge of a primary control surface. Tabs may be used for:

- Control balancing, to assist the pilot to move the control
- Servo operation of the control
- Trimming

The balance tab (Figure 7.66) is used to reduce the hinge moment produced by the control and is therefore a form of aerodynamic balance, which reduces the effort the pilot needs to apply to move the control.

The tab arrangement described above may be reversed to form an anti-balance tab, as seen in Figure 7.67. The anti-balance tab is connected in such a way as to move in the same direction as the control surface, so increasing the control-column loads. This tab arrangement is used to give the pilot feeling, so that the aircraft will not be overstressed as a result of excessive movement of the control surface by the pilot.

The spring tab arrangement (Figure 7.68) is such that tab movement is proportional to the load on the control, rather than the control-surface deflection angle. Spring tabs are used mainly to reduce control loads at high speeds. The spring is arranged so that below a certain speed it is ineffective – Figure 7.68(a). The aerodynamic loads are such that they are not sufficient to overcome the spring force and the tab remains in line with the primary control surface. As speed is increased, the aerodynamic load acting on the tab is increased sufficiently to overcome the spring force and the tab moves in the opposite direction to the primary control to provide assistance, Figure 7.68(b).

The servo tab is designed to operate the primary control surface – Figure 7.69. Any deflection of the

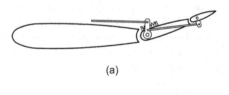

(a)

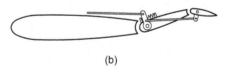

(b)

7.68 Spring tab arrangement

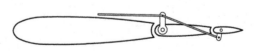

7.69 Servo tab

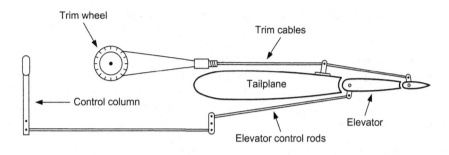

(a)

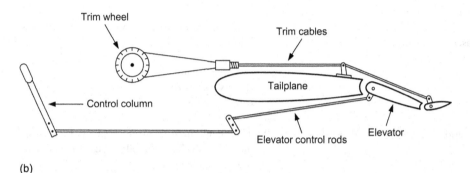

(b)

7.70 Typical manual elevator trim system

tab produces an opposite movement of the free-floating primary control surface, thereby reducing the effort the pilot has to apply to fly the aircraft.

The trim tab is used to relieve the pilot of any sustained control force that may be needed to maintain level flight. These out-of-balance forces may occur as a result of fuel use, variable freight and passenger loadings or out-of-balance thrust production from the aircraft engines. A typical manual elevator trim system is illustrated in Figure 7.70(a). In this example, the pilot rotates a trim wheel, which provides instinctive movement. If the pilot rotates the elevator trim wheel forward, the nose of the aircraft will drop; the control column will also be trimmed forward of the neutral position, as shown in Figure 7.70(b).

Under these circumstances the trim tab moves up, the elevator is moved down by the action of the trim tab, thus the tail of the aircraft will rise and the nose will pitch down. At the same time the elevator control rods move in such a way as to pivot the control column forward. A similar set up may be used for aileron trim, where in this case the trim wheel would be mounted parallel to the aircraft lateral axis and

rotation of the wheel clockwise would drop the starboard wing, again the movement being instinctive.

MULTIPLE-CHOICE QUESTIONS

1. The angle of attack is the angle between the:

   (a) Chord line and the relative airflow
   (b) Relative airflow and the longitudinal axis of the aircraft
   (c) Maximum camber line and the relative airflow

2. If the angle of attack of an aerofoil is increased, the centre of pressure will:

   (a) Move backward
   (b) Stay the same
   (c) Move forward

3. The angle of incidence on conventional aeroplanes:

   (a) Varies with aircraft attitude
   (b) Is a predetermined rigging angle
   (c) Is altered using the tailplane

4. Interference drag may be reduced by:

   (a) Fairings at junctions between the fuselage and wings
   (b) Highly polished surface finish
   (c) High-aspect-ratio wings

5. If lift increases, vortex drag:

   (a) Decreases
   (b) Remains the same
   (c) Increases

6. The aspect ratio of a wing may be defined as:

   (a) Chord/span
   (b) Span squared/area
   (c) Span squared/chord

7. In a climb at steady speed the thrust is:

   (a) Equal to the drag
   (b) Greater than the drag
   (c) Less than the drag

8. Movement of an aircraft about its normal axis is called:

   (a) Rolling
   (b) Pitching
   (c) Yawing

9. The dimension from wing tip to wing tip is known as:

   (a) Wingspan
   (b) Wing chord
   (c) Aspect ratio

10. The device used to produce steady flight conditions and relieve the pilot's sustained control inputs is called a:

    (a) Balance tab
    (b) Trim tab
    (c) Servo tab

8 Aeroplane automatic flight control

The technology used for the automatic control of aeroplane autopilots varies in complexity depending on the intended functionality. The term 'autopilot' refers to a range of technology encompassing the simple wings leveller, through to fully automatic approach and landing. The functionality of an autopilot in its basic mode of operation is to maintain the aircraft on its desired flight path. Autopilots can also be integrated with a flight-director system; in this case the commands are presented on the flight instruments allowing the pilot to follow a desired flight path. Flight directors can be operated separately for manual control of the aircraft, or (with the autopilot engaged) allowing the pilot to monitor the autopilot's performance. The main purpose of an autopilot/flight-director system is to relieve the pilot of the physical and mental fatigue of flying the aircraft. This is particularly beneficial on long-flight legs, or in low-visibility operations. Aviation medical research and flight testing has established that an automatic system is more responsive than a human pilot. There are many types of autopilot in operational use, ranging from simple analogue systems used on general aviation aircraft through complex digital systems used on large commercial transport aircraft. This chapter establishes the basic principles based on practical examples; the schematics used are based on analogue systems, to illustrate basic principles. Combined autopilots and flight directors are referred to by various names, depending on the technology and manufacturer, e.g. automatic flight-control system, automatic flight-guidance system, digital flight-guidance system.

PRINCIPLES

All autopilot and flight-director systems are based on four servo-based system principles:

- Error-sensing input(s)
- Correction
- Feedback
- Commanded output(s)

The error-sensing input is basically a disturbance detector, based on an attitude reference sensor, typically a vertical gyroscope. The attitude sensor will sense if the aircraft is deviating from the desired flight path. Referring to Figure 8.1, datum angles are

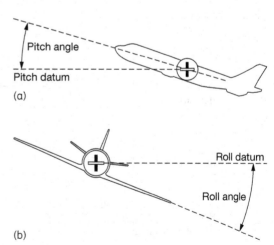

8.1 Datum angles: (a) Pitch (b) Roll

established by the vertical gyroscope; this can either be a dedicated sensor, or derived from an attitude-reference system. Deviations from straight and level flight are sensed in pitch and/or roll. Deviation(s) is referred to as an error signal; this is amplified and used to drive a motor that moves a control surface, as seen in Figure 8.2. With reference to the earlier chapter on control systems, some feedback is required from the motor to optimize the system's response. The servo-motor's position (and on larger aircraft the control surface position) is fed back into the correction system. In some systems, rate feedback is also used to enhance system response. Commanded outputs are based on a variety of external references, e.g. heading hold and altitude hold. The flight director will indicate the required control inputs that the pilot follows; the autopilot will move the control surfaces via servo motors. Combined autopilot/flight-director systems are often referred to as automatic flight-control systems (AFCS) or automatic flight-guidance systems (AFGS).

KEY POINT

A servo motor provides position and/or rate feedback into a control system.

Typical autopilots control the aircraft in three axes, or channels: pitch, roll and yaw (see Figure 8.3). Each axis, or channel, can be controlled via various commanded outputs; these are described here as separate control axes; in reality they all act simultaneously. Smaller general aviation autopilots are based on one or two axes of control: roll only or pitch and roll. In the very basic systems, roll control provides a wings-level function.

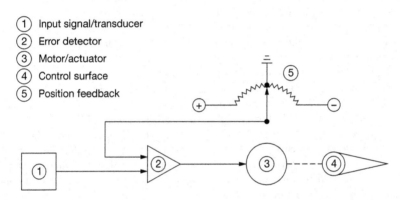

1. Input signal/transducer
2. Error detector
3. Motor/actuator
4. Control surface
5. Position feedback

8.2 Servo principles

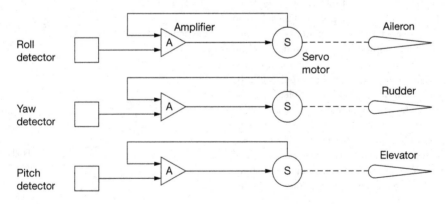

8.3 Three-axis control

Autopilots can be based on attitude-based displacement gyroscopes or rate gyroscopes. There are various considerations in both types, including reliability and performance; these are more significant for general aviation and/or older electromechanical systems. Higher specified systems on modern aircraft tend to be based on solid-state, strapdown technology. Electromechanical rate gyros operate at much lower speeds compared with displacement gyroscopes, typically 50 to 70 per cent lower, meaning that the mean time between failure (MTBF) is much lower on rate-based gyroscopes. The bearings in all electromechanical gyros will eventually degrade over time. In displacement gyroscopes, these will lead to precession, which ultimately affects performance. Rate-based gyros can maintain performance even with worn bearings. Finally, rate gyroscopes do not suffer from tumble at unusual attitudes.

PITCH CONTROL

Referring to Figure 8.4, consider an aircraft that is flying straight and level, and is then disturbed in pitch – nose down. The vertical gyro (VG) senses the change in attitude and produces an error signal. This error signal depends on the system design; the typical arrangement is in the form of a synchro system. The error signal produces an 'up-elevator' output and this is processed in the autopilot's pitch computer, as seen in Figure 8.5. The pitch computer processes and amplifies the error signal and then drives the servo motor to move the control surface. To achieve optimum response from the control system, feedback is produced; this limits the amount (and in some systems the rate) of corrective 'up-elevator' control.

These feedback signals are typically produced by synchro transmitters identified as TX1 and TX2 in the

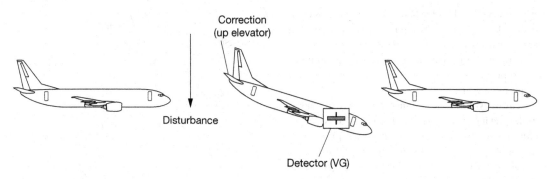

8.4 Aircraft disturbed in pitch

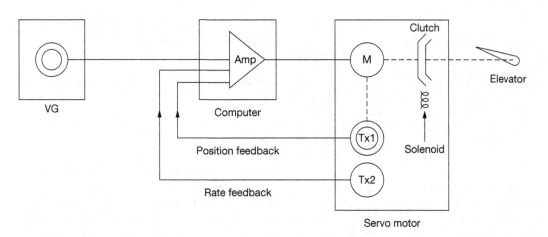

8.5 Error signal producing an 'up-elevator' output

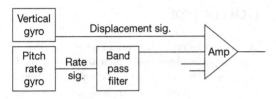

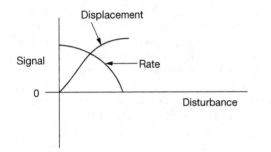

8.6 Rate/displacement response

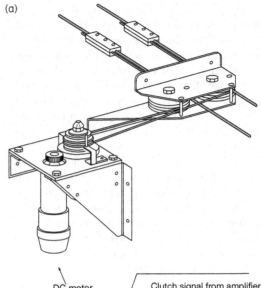

(a)

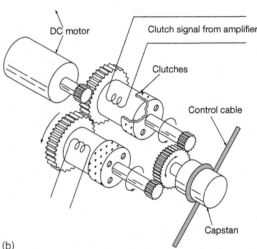

(b)

8.7 Autopilot servomotor: (a) Servomotor and control cables (b) Servomotor and clutch arrangements

system diagram. This feedback is referred to as negative feedback since it is in opposition to the originating error signal. Autopilots can either be based on displacement or rate-based sensors, i.e. depending on the type of gyroscope, or derived signal; typical responses are given in Figure 8.6.

The servomotor is connected into the control cables and incorporates a clutch – see Figure 8.7. This clutch is used to engage/disengage the pitch servo motor. When the pilot is flying the aircraft manually, the autopilot is off and the clutch disengaged. When the autopilot is engaged, power is applied to the clutch via a solenoid, and the clutch connects the motor to the control cables.

The applied 'up-elevator' not only prevents any further 'nose-down' attitude change, but also takes the aircraft back to its original pitch attitude. As the VG error signal reduces to zero the predominant position feedback signals causes opposite servo-motor rotation

to remove the control surface deflection and its signal to zero.

Damping

To augment stabilization of the aircraft's flight path, rate gyro signals can also be incorporated into the control system. With a VG only, as described above, the error signal builds as the disturbance increases. By introducing a rate signal, its error signal is at maximum when the disturbance starts and zero when the

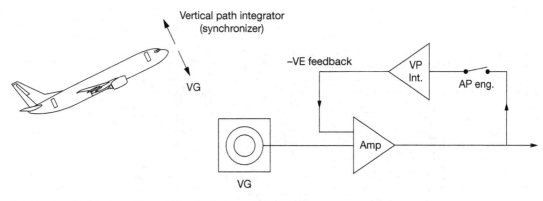

8.8 Synchronizing to the existing pitch attitude

disturbance reaches a steady state. A band pass filter ensures that only the aircraft's natural frequency of oscillation is processed.

Synchronization

Prior to autopilot engagement, the pitch channel is automatically synchronized to the existing pitch attitude of the aircraft (Figure 8.8). This prevents transients from adversely acting on the aircraft when the autopilot is engaged; the existing aircraft attitude is maintained upon engagement. In this example, the synchronization is achieved by a vertical path integrator, referred to as a 'washout' function. Most autopilots incorporate a pitch trim control allowing the pilot to manually control the desired aircraft attitude.

Manometric commands

Most autopilots incorporate one or more pitch command functions based on manometric modes of pitch command:

- Altitude hold
- Vertical speed
- Altitude select
- Level change
- Indicated airspeed (IAS) hold

The desired mode of operation is selected from the autopilot's control panel; the selected mode of operation is displayed via annunciators. Typical mode-select panels (MSP) are shown in Figure 8.9. Manometric inputs to the autopilot are derived from the pitot static system, e.g. a centralized air-data computer (CADC).

When the altitude-hold mode is engaged via an ALT HOLD selector switch on the MSP – Figure 8.10 – a clutch is engaged in the CADC enabling an altitude error signal to be supplied to the pitch computer (PC). When this mode is engaged, the aircraft will maintain the existing barometric altitude. Any deviation in altitude will produce an error signal from the air-data computer's altitude module. The gain of the PC is normally varied as a function of airspeed; the commanded output will have limited authority, typically $+/-7.5$ degrees of control surface movement.

Indicated airspeed (IAS) hold operates in a similar way, with the PC referenced to an airspeed module in the CADC. Pitch control of the aircraft is maintained at the existing IAS when the mode is engaged. Some aircraft have a variable IAS function whereby the pilot selects the desired IAS on the mode-selector panel. A vertical speed mode controls the aircraft pitch attitude via an error signal. Engaging the VS mode will maintain the aircraft's vertical speed by controlling the aircraft's pitch attitude.

The altitude-select mode is typically based on two altitude error output signals from the CADC: coarse and fine. Referring to Figure 8.11, altitude select mode operates in a succession of three phases:

- Arm
- Capture
- Hold

(a)

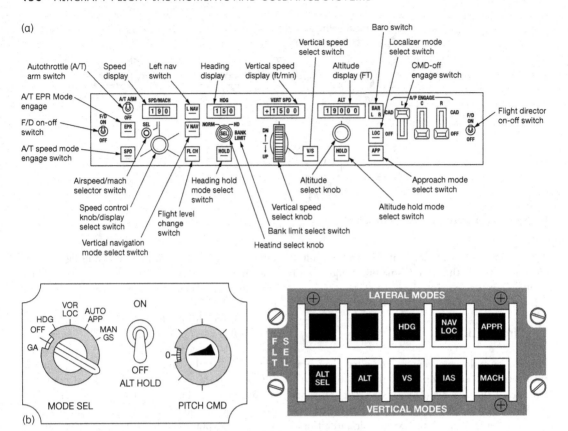

8.9 Typical mode-select panel (MSP): (a) Transport aircraft (b) General aviation

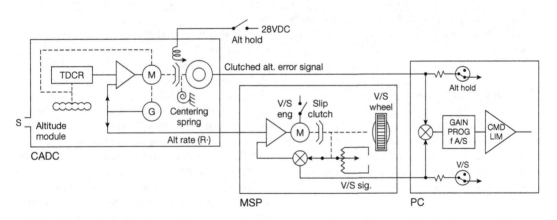

8.10 Altitude-hold mode

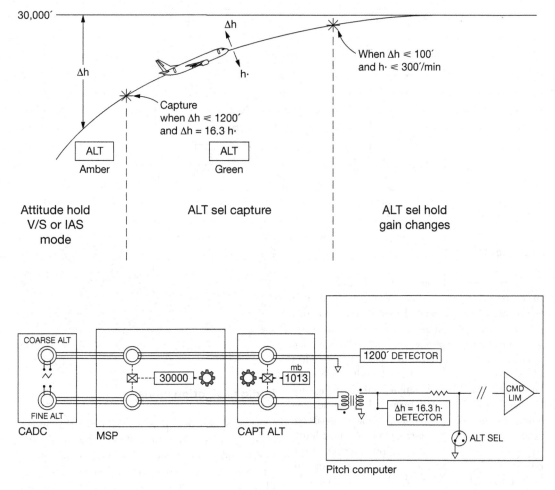

8.11 Altitude-select mode

The desired altitude, e.g. 30,000 feet, is selected on the mode-select panel (MSP). During the altitude-select arm phase, the flight path used to intercept the selected altitude is established by either a pilot-selected attitude, VS or IAS modes. When the two detectors in the pitch computer sense that the aircraft is within a predetermined altitude (typically 1,000 feet) of the selected altitude and closure rate (typically 10 feet / second), the capture phase begins. The previous flight-path mode (either selected attitude, VS or IAS) is automatically disengaged. The altitude-hold phase is achieved when the difference between actual altitude and selected altitude is typically 100 feet and altitude rate is typically less than 5 feet / second.

ROLL CONTROL

The basic autopilot operation for roll stability and control is similar to that described for pitch. Attitude is sensed from an electromechanical vertical gyro or solid-state attitude reference system; most autopilots also have a manual control feature for pitch and roll, as seen in Figure 8.12. Systems designed with rate sensors can derive this by differentiating the attitude signal, or by a rate gyro. Typical roll command modes include:

- Heading hold
- Heading select
- Radio navigation
- Area (or lateral) navigation

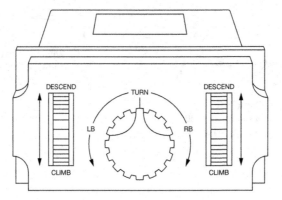

8.12 Manual pitch/turn controller

Heading-hold mode

This is the basic roll-command mode and provides automatic hold to the heading existing at the time of engagement. Most autopilots have a design feature that prevents heading hold if the roll attitude exceeds a predetermined value, typically three degrees. Operation of a turn-control knob can be used at any time to select a bank angle to achieve the desired heading. When the knob is returned to the detent, and the wings are levelled (within the predetermined value) heading hold is automatically initiated. Command signals are programmed with predetermined bank-angle limits (typically +/− 30 degrees) and predetermined roll rate (typically 4 degrees/second). These authority limits can be scheduled in accordance with airspeed.

Heading-select mode

The heading signal is derived from the aircraft's compass system, typically the attitude heading reference system (AHRS). An error signal is derived from the aircraft's actual heading and the heading selected on the mode-select panel. Roll-rate authority is limited; some systems have a variable-rate limit ranging from 1 to 3 degrees per second to allow for large heading changes to be made depending on airspeed. Rate limit can also be adapted when heading select is used in conjunction with other roll modes.

TEST YOUR UNDERSTANDING 8.1

An autopilot that controls the aircraft in altitude, heading and roll attitude is a two- or three-axis system?

Radio navigation modes

Most autopilots incorporate a roll mode that allows capture and tracking of VHF Omnirange (VOR) radials and localizer course. The mode-select panel will have a combined VOR/LOC switch position. (The localizer function is covered under the subject of instrument landing systems.) This mode consists of four phases:

- Intercept
- Capture
- On-course
- Over-station

In the first instance, VOR mode requires that (i) VOR frequency is selected on the VHF navigation control panel, (ii) a VOR radial is selected on the mode-select panel and (iii) VOR is selected on the autopilot's mode-select panel. Referring to Figure 8.13, in the intercept phase, heading-select or turn-control modes will be used to steer the aircraft to the selected radial. When aircraft approaches a predetermined angle from the selected radial, shown here as 10 degrees, this will generate an error signal of 150mV from the

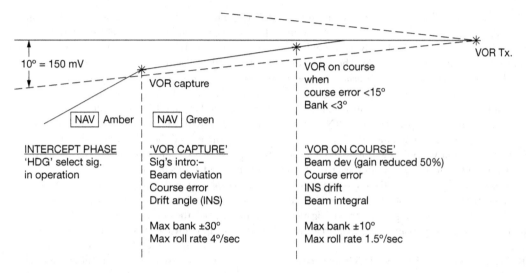

8.13 VOR mode

navigation receiver. This is the VOR capture angle, heading select is disengaged; the autopilot's error signals are now beam deviation and course error angle. Maximum bank angle and roll rates are now in place, typically +/− 30 degrees and 4 degrees/ second.

An on-course (O/C) sensor in the roll computer is tripped at predetermined levels of bank angle (typically less than three degrees) and course error (typically less than 15 degrees). Bank-angle and roll-rate limits are reduced again, typically +/− 10 degrees and 1.5 degrees/second. As the VOR station

is approached, the beam-rate signal increases because of the beam distortion overhead of the station (sometimes called the 'cone of confusion') – see Figure 8.14. When the beam-rate signal increases to approximately +/− 8 mV/second overhead of the station, beam deviation gain is reduced by 50 per cent.

Area (lateral) navigation modes

Area navigation (RNAV) is a means of combining, or filtering, inputs from one or more navigation sensors and defining positions that are not necessarily co-located with ground-based navigation aids. This facilitates aircraft navigation along any desired flight path within range of navigation aids; alternatively, a flight path can be planned with autonomous navigation equipment. RNAV is used for long-range, en-route navigation where there are few, if any, radio navigation aids.

Optimum RNAV can be achieved using a combination of ground navigation aids and autonomous (on board) navigation equipment. Typical navigation sensor inputs to an RNAV system can be from external ground-based navigation aids such as VHF omnirange (VOR) and distance-measuring equipment (DME); autonomous systems include global positioning satellite (GPS) navigation or inertial-reference system (IRS). Many RNAV systems use a combination of numerous ground-based navigation aids, satellite-

HSI indications:
 – Bearing pointer removed
 – CDI removed
 – Numeric bearing removed

VOR cone-of-confusion

Station passage:
 – Bearing pointer, CDI, and other
  data reappear exiting the cone

8.14 VOR cone of confusion

navigation systems and self-contained navigation systems. In this chapter, we will focus on area navigation systems that use VOR and DME navigation aids to establish the basic principles of RNAV. This subject is covered in more detail in *Aircraft Communications and Navigation Systems (ACNS)*, which includes a review of Kalman filters and how RNAV systems are specified with a 'required navigation performance' (RNP). Area navigation systems contain a database of pre-stored flight plans based on waypoints and individual flight legs (or tracks) that form a route. In this chapter, an example is given of how the inertial reference system (IRS) or inertial navigation system (INS) is coupled to the autopilot. The IRS is a long-range area navigation system; the same principles apply to other area navigation systems.

Referring to Figure 8.15, when INS is selected on the mode-select panel, the autopilot will capture and maintain any desired track in the flight plan. This mode typically contains four phases:

- Arm
- Capture
- Track
- Waypoint switching

When the aircraft is beyond a pre-set distance, typically 7.5nm from the desired track, the roll channel is in the arm phase with heading select used to steer the aircraft. A cross-track deviation signal from the INS is applied to the capture sensor; when the 7.5nm threshold is reached, the capture phase is initiated. INS cross-track deviation (CTD) and track-angle error (TAE) are filtered and summed in the roll computer; the resultant signal commands the aircraft to fly at a fixed intercept angle to capture the desired track. When the CTD and TAE are reduced to pre-set values, typically 1000 feet/3 degrees, the INS course sensor is tripped and the roll channel switches to the track phase. Bank angle and roll-rate limits are normally reduced at this point.

Geographical positions defined in an RNAV system are called waypoints. Automatic waypoint switching is initiated when the aircraft is in the INS track phase. Depending upon the angle between the present track and the new desired track, a signal is produced by the INS at a fixed distance from the new track. The capture phase is recycled by this signal to acquire the new desired track. The waypoint is actually bypassed during this transition, unless the INS (or other area navigation system) is specifically programmed to fly overhead of the waypoint.

Instrument landing system

Navigation aids such as automatic direction finder (ADF), VHF omnidirectional range (VOR) and distance-measuring equipment (DME) are used to define airways for en-route navigation. They are also

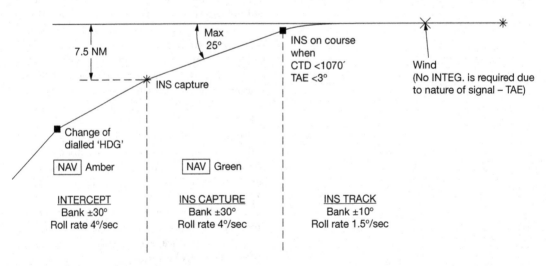

8.15 Long-range navigation lateral mode

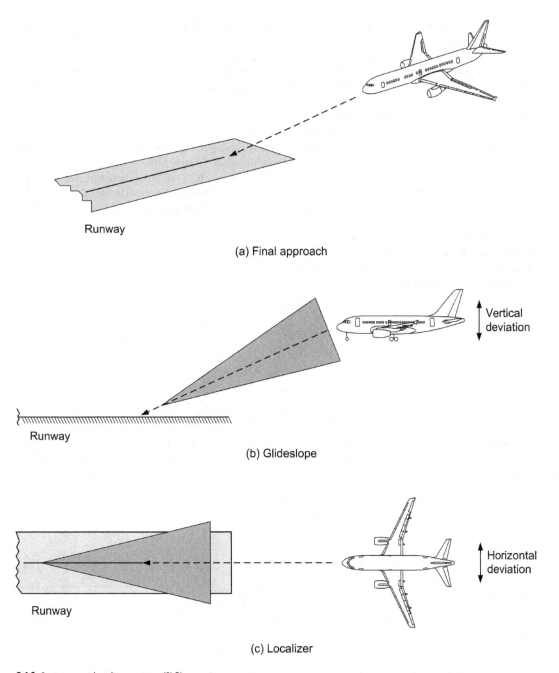

(a) Final approach

(b) Glideslope

(c) Localizer

8.16 Instrument landing system (ILS) overview

installed at airfields to assist with approaches to those airfields. These navigation aids cannot, however, be used for precision approaches and landings. The standard approach-and-landing system installed at airfields around the world is the instrument landing system (ILS).

The instrument landing system is used for the final approach and is based on directional beams propagated from two transmitters at the airfield – see Figure 8.16. One transmitter (the glide slope) provides guidance in the vertical plane and has a range of approximately 10nm. The second transmitter (the localizer) guides the aircraft in the horizontal plane. In addition to the directional beams, two or three marker beacons are located at key points on the extended runway centreline defined by the localizer (Figure 8.17a). These marker beacons are identified on a display to the pilot (Figure 8.17b).

Localizer (LOC) mode

This is a very similar mode to VOR; most autopilot control panels have a single selector switch for both modes, typically identified as VOR/LOC or VOR/L. Some autopilot control panels have these functions combined with ILS or LAND (see below in this chapter).

Referring to Figure 8.18, when VOR/L is selected on the mode-control panel, and a localizer frequency is tuned by the VHF navigation receiver, the roll channel automatically operates in the LOC mode. The aircraft is controlled to capture and track the localizer beam in three phases:

- Arm
- Capture
- On-course

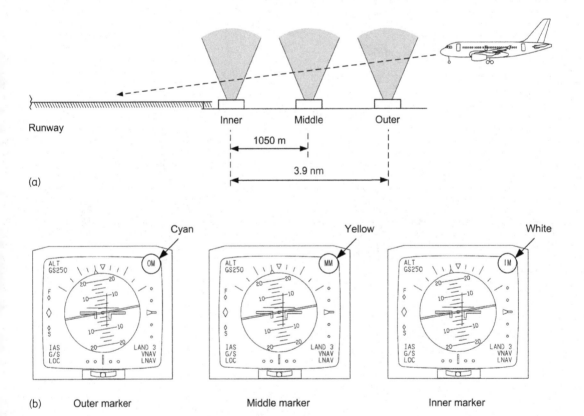

(a)

(b)   Outer marker   Middle marker   Inner marker

8.17 ILS marker beacon overview: (a) Location on approach path (b) Marker beacon displays

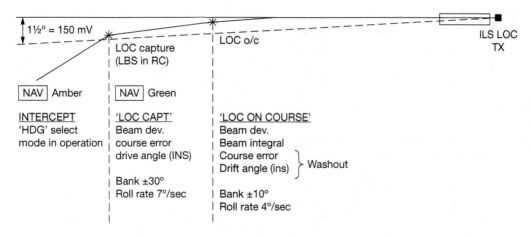

8.18 ILS localizer (LOC) mode

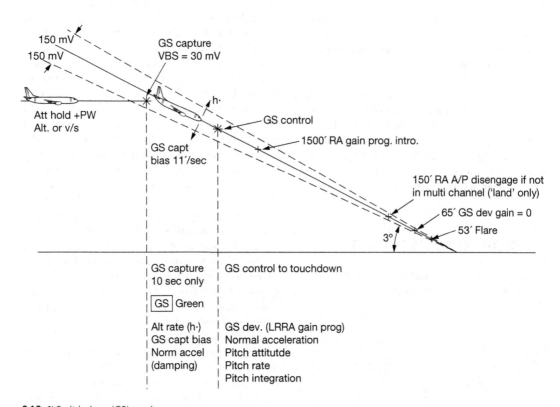

8.19 ILS glideslope (GS) mode

The aircraft is steered to the localizer in another mode, e.g. area navigation or heading. The VOR/LOC capture lateral beam sensor (LBS) will be in receipt of course error, beam error and beam-rate signals. At a pre-set threshold, the LBS transfers roll control from the previous steering mode to LOC capture. Beam deviation course error and (where installed) drift angle from an area navigation system, e.g. GNSS. Roll rate limits are now reduced to a lower value, typically 7 degrees/second. When the beam deviation is less than 60 mV, beam rate is less than 2 mV/second and bank angle is less than 3 degrees, then LOC on-course begins. Maximum bank angle is now +/− 10 degrees and beam rate is limited to 4 degrees/second. From 1,500 feet radio altitude, the roll computer's localizer gain reduces the beam error gain as a function of radio altitude to compensate for narrowing of the localizer beam.

Pitch coupling to the ILS glide slope (GS) – Figure 8.19 – is achieved when the vertical beam sensor (VBS) in the pitch computer senses that the glide slope deviation signal has reduced to 30mV. The glide slope is normally intercepted from another pitch command mode, e.g. altitude hold. Once the glide slope is captured, altitude hold is disconnected (now being incompatible) and a nose-down bias signal is injected by the pitch computer to cause the aircraft to pitch down by a controlled amount. This nose-down command is sometimes blended with other pitch sensors to smooth the transition, e.g. altitude rate, pitch angle, pitch rate, normal acceleration.

After a pre-set time period, typically 10 seconds, the GS control phase begins and any other pitch sensors are blended into the control function. From a pre-set radio altitude, typically 1,500 feet, GS deviation signal gain is gradually reduced to account for narrowing of the GS beam. At a typical radio altitude of 150 feet, the autopilot disengages, unless it is part of an autoland system (see Chapter 11).

TEST YOUR UNDERSTANDING

Explain why altitude hold is incompatible with the glide-slope tracking mode.

KEY POINT

The glide slope would normally be captured from the altitude hold mode, from below the GS.

KEY POINT

Using vertical speed mode from above to capture the GS is possible, although it is not the preferred method for most autopilots.

YAW DAMPERS

Some autopilot systems have a dedicated yaw damper system; in some aircraft the yaw damping function is incorporated into the roll and/or rudder channel computer. For ease of explanation, this section considers a dedicated yaw damper system.

All high-speed aircraft are, to a certain extent, subject to a phenomenon known as Dutch roll, as described above. This situation occurs with a combined yaw and roll disturbance and is sometimes referred to as oscillatory stability. When an aircraft is disturbed such that it yaws, this also causes a secondary disturbance in the roll axis. The fin and rudder are used to compensate the disturbance and recover stability. In some aircraft at certain speeds, the disturbance cannot be adequately corrected, resulting in yaw instability (Figure 8.20). As with other forms of stability, this can be positive, neutral or negative.

A practical yaw damper system is based on a directional rate gyro. The gyro feeds a signal to the yaw damper which then applies a corrective signal to the rudder servomotor. The rudder displacement opposes the yaw disturbance to correct the disturbance, normally within two to three cycles of the Dutch roll. If the disturbance occurs with the yaw damper switched off, then engaging the system will bring the aircraft under control (Figure 8.21).

The Dutch roll frequency of an aircraft is based on its natural yawing frequency, typically 0.2/0.4Hz on medium/large transport aircraft. Yaw dampers are designed to allow the Dutch roll frequency to demand rudder in opposition to the Dutch roll yawing disturbance, but must block other frequencies (Figure 8.22) to prevent the rudder opposing normal aircraft turns. A yaw damper filter based on an analogue filter is used to illustrate this function – see Figure 8.23 (a). This is a simplified circuit, used to illustrate the principles of the system. This narrow-band filter is designed to pass only the signals that change at the Dutch roll

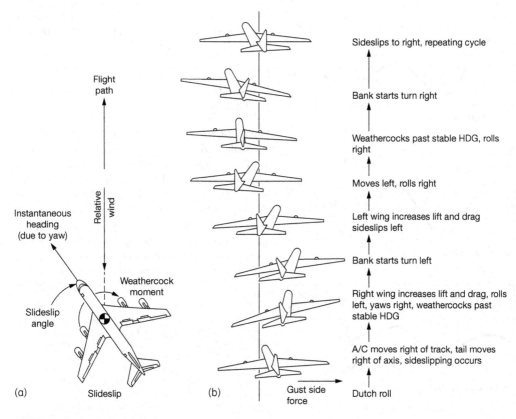

(a) Slideslip

Flight
path

Relative
wind

Instantaneous
heading
(due to yaw)

Weathercock
moment

Slideslip
angle

(a)   Slideslip

(b)   Gust side force   Dutch roll

Sideslips to right, repeating cycle

Bank starts turn right

Weathercocks past stable HDG, rolls right

Moves left, rolls right

Left wing increases lift and drag sideslips left

Bank starts turn left

Right wing increases lift and drag, rolls left, yaws right, weathercocks past stable HDG

A/C moves right of track, tail moves right of axis, sideslipping occurs

8.20 (a) Side-slip (b) Gust side force

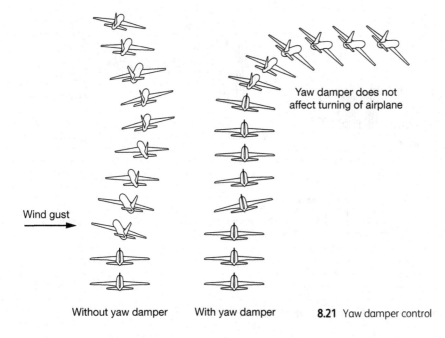

Wind gust

Without yaw damper   With yaw damper

Yaw damper does not affect turning of airplane

8.21 Yaw damper control

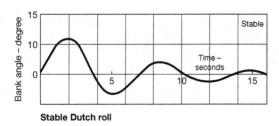

Stable Dutch roll

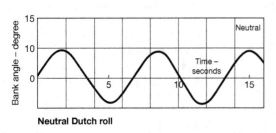

Neutral Dutch roll

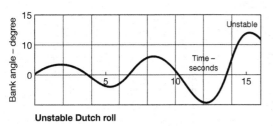

Unstable Dutch roll

8.22 Dutch roll characteristics

frequency. The rate gyro produces outputs for all turns, but only those related to Dutch roll will be sent to the rudder servomotor.

1. Stable
2. Neutral
3. Unstable

As illustrated in Figure 8.23(b), when the rate of turn is initially increasing up to a constant rate, the Dutch roll filter output also builds, then reduces to zero when the rate of turn becomes constant. The reverse occurs with opposite polarity as the filter's capacitor discharges; the aircraft levels out on completion of the turn, resulting in no rudder demand. In Figure 8.23(c), the aircraft is yawing at the Dutch roll frequency, since the rate of turn is constantly changing; the output from the rate gyro is constantly changing. The DC graph illustrates the Dutch roll filter's output.

> **KEY POINT**
>
> The DC polarities of the Dutch roll filter's outputs are the greatest when the rate of turn is maximum, and zero when the rate of turn is zero.

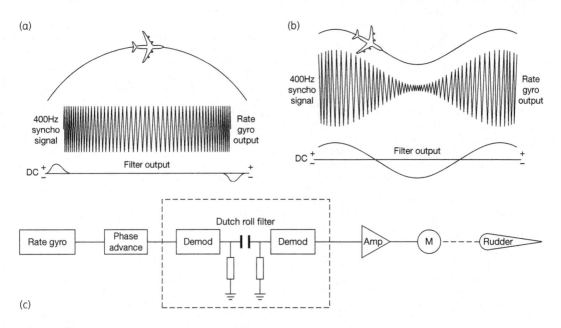

8.23 Yaw damper frequency analysis: (a) Constant turn (b) Dutch roll (c) Filtering

CONTROL LAWS

At the beginning of this chapter, basic autopilot theory was described using the concepts of error-sensing input(s), correction, feedback and commanded output(s). Different aircraft types have differing aero-dynamic characteristics so, for a given disturbance, will require different amounts of correction and feedback for basic stability. In the subsequent sections of this chapter, examples have been given of varying roll/pitch limits and rates depending on the particular phase of automatic flight.

One way of varying the rate of correction to a given disturbance is by altering the feedback signal. In a typical analogue-based autopilot system, this is achieved by altering the value of pre-set resistors; in a digital system by invoking different software routines. This enables the feedback signal level to be adjusted and so permit changes in the servomotor gearing ratio.

The pitch attitude of an aircraft is controlled by the elevators; the directional control is from both rudder and ailerons. At higher speeds, directional control is more effectively achieved from the ailerons rather than the rudder. The roll-channel computer will have a speed input from one of the aircraft sensors, e.g. the air data computer, that will be used to alter the gain of signals used to control the ailerons and rudder. The efficiency of the autopilot in correcting yaw distur-bances will be increased by controlling the rudder and ailerons simultaneously. This is achieved by applying directional gyroscope displacement and/or rate signals into the roll computer and (where installed) yaw damper. This arrangement is termed crossfeed, and is used to coordinate turns made for small deviations from the desired heading, or track.

Some autopilots incorporate a turbulence mode where pitch-gain circuits are reduced by up to 50 per cent to provide pitch-attitude hold function that is stabilized with softened commands to reduce loads on the airframe and to smooth the flight path for the convenience of passengers.

INTERLOCKS

Before the autopilot is engaged, and control of the aeroplane transferred to automatic, it is important that a number of conditions are satisfied to ensure no hazards are introduced. For example, the trim of the aeroplane must be established by the pilot to avoid a sudden attitude change when the autopilot is engaged. Similarly, all power supplies and sensors to the autopilot system, together with system outputs, must be operational and within limits. The interlocks are designed to prevent engagement of the autopilot until all requirements are satisfied.

The automatic pilot integrates signals from various sensors and auxiliary controls; positive sequencing of engagement is required to prevent improper opera-tion. Protection also has to be made against adverse interaction of integrated components, resulting from a malfunction. Various components, e.g. switches and relays, are connected in a specific way, referred to as interlocking, to provide this sequencing, control and protection.

The scope of interlocking depends on the com-plexity of the system and its intended function, e.g. a simple wings-leveller system will check for an attitude-reference signal and electrical power to the servo motor. Multiple-mode systems will also ensure that incompatible modes are not engaged at the same time, e.g. altitude hold/vertical speed; heading hold/ILS.

Various interlocks must be satisfied before the autopilot ON/OFF switch will latch in the engaged position.

KEY POINT

Interlocks prevent the automatic flight-control system from being powered up until the system is capable of controlling the aircraft.

CASE STUDY – GENERAL AVIATION AUTOPILOT

An autopilot system developed for general aviation aeroplanes is Avidyne's DFC90/100 product range. This system is designed for retrofit on GA aircraft, and is described here to illustrate typical autopilot functionality.

The DFC90/100 is an autopilot control panel (ACP) and autopilot computer integrated into one single unit (Figure 8.24). The front bezel of the

8.24 DFC90/100

DFC90/100 allows the flight crew to select the various flight modes. The DFC90/100 computer receives information from the aircraft's navigation and flight instruments to provide two-axis (pitch and roll)

control of the aircraft. The DFC90/100 autopilot provides the following flight functions for piloting the aircraft:

- Straight & Level
- Heading-mode navigation (NAV, APPR)
- Roll steering (GPSS) mode (DFC90 Only)
- Altitude capture and hold mode
- Vertical-speed mode
- Indicated-airspeed mode
- Glideslope (GS) Mode
- VNAV mode (DFC100 Only)
- Control-wheel steering

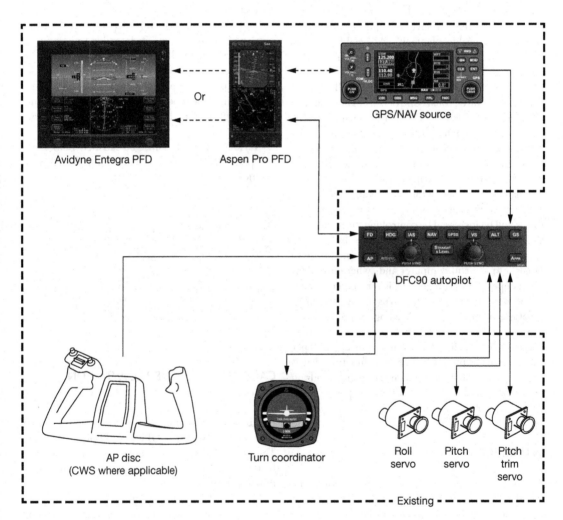

8.25 DFC block diagram

Figure 8.25 depicts a typical system block diagram for (a) the DFC90 autopilot and (b) the DFC100. The Avidyne autopilot is an attitude-based autopilot that provides dual-axis flight control in a Piper PA-46 aircraft. The DFC90/100 receives data from several external sources and outputs pitch-and-roll steering commands to the servo/trim motors. The autopilot is also interfaced to the aircraft's electric pitch-trim system to provide automatic pitch trim of the aircraft. The autopilot will automatically trim the aircraft when the autopilot is engaged in any vertical flight mode. If an optional yaw-damper system is installed, the autopilot will activate the yaw-damper system when any lateral flight mode is selected. The autopilot controller is normally located in the centre instrument panel as shown in Figure 8.26.

The DFC90 will utilize either an Entegra EXP5000 or Aspen EFD1000 Pro PFD as the attitude source. The DFC90 adds the precision of an attitude-based

flight-control system and Avidyne's innovative, safety-enhancing EP™ capability.

The DFC90 has all the standard vertical and lateral modes of operation of a turbine-class autopilot system, including flight director (FD), altitude hold (ALT), airspeed hold (IAS), vertical-speed hold (VS), heading (HDG) and navigation (NAV, APPR, LOC/GS, GPSS).

The attitude-based design improves stability due to the use of attitude data to control the autopilot inner control loops. This is particularly evident and important when tracking an ILS to minimums in windy conditions. The 'straight & level' function recovers the aircraft from unusual attitudes. This overrides all autopilot modes and levels the aircraft in both pitch and roll from a wide range of capture attitudes for an added measure of safety.

The indicated-airspeed hold (IAS) includes a dedicated airspeed knob and a new airspeed bug on

8.26 DFC installation

the PFD. It provides constant speed as flight level changes. The synchronized heading bug enables turns that are commanded in correlation with the heading knob. The autopilot does a right-hand 270 degree turn if the pilot turns the heading knob right 270 degrees.

Envelope Protection (EP™) prevents autopilot-induced stalls and over-speeds – which have previously been a major contributing factor in GA accidents – when the autopilot is engaged. EP also provides visual and aural warnings to the pilot. With Avidyne's Envelope Alerting EA™, available lift and speed margins are calculated constantly in the background, providing visual and aural warnings to the pilot when the aircraft is nearing its normal flight parameters, even when the autopilot is off.

Aural alerts

Aural alerting, through the aircraft intercom system, is provided for warnings from the autopilot. In the context of the bullets below, 'coupled' describes the condition when the autopilot servos are flying the airplane and 'non-coupled' describes the condition when the servos are not flying the airplane and instead the pilot is expected to follow the flight-director command bars, if present. Specifically, aural alerts as defined in the parentheses are provided under the following conditions:

• Autopilot disengaged (approximately 16 disconnect beeps)
• Underspeed during coupled operations ('Speed-protection active')
• Overspeed during coupled operations ('Speed-protection active')
• Underspeed during non-coupled operations ('Caution, underspeed')
• Overspeed during non-coupled operations ('Caution, overspeed')
• Attitude and heading-reference system (AHRS) and
• Turn coordinator miscompare ('Gyro miscompare')
• Bank limit exceeded ('Caution, excessive bank')
• Flap limit exceeded ('Caution, flap overspeed')

Autopilot engagement

From a standby state (autopilot has power, 'AP READY' displayed but no modes are engaged and the airplane is within the engagement limits defined above), pressing any button on the autopilot (except GS and APPR) will engage the DFC autopilot. If a specific lateral and/or vertical mode is not pressed, the system will default to ROLL hold mode in the lateral channel and PITCH hold mode in the vertical channel. From a standby state, the pilot selects:

• 'AP' – autopilot (servos coupled) engages in ROLL and PITCH and will hold whatever bank and pitch was present at time of pressing (assuming within command limits).
• 'FD' – flight director (servos not coupled) engages in ROLL and PITCH and will command via the green flight-director command bars whatever bank and pitch was present at time of pressing (assuming within command limits). It is still up to the pilot to manoeuvre the plane as required to follow those command bars.
• 'STRAIGHT & LEVEL' autopilot (servos coupled) engages and drives the airplane from whatever attitude it is in to zero bank and a small positive pitch that approximates level flight.
• Any other button(s) (except GS or APPR) on the autopilot (servos coupled) engages and will enter the modes as commanded.

Autopilot disengagement

The autopilot can be disengaged at any time using any one of the following methods:

• Pressing the AP disconnect switch on the control yoke
• Activating the pitch-axis trim switch on the control yoke
• Pressing the 'AP' button on the autopilot control panel (servos will disconnect but the flight director will remain active)
• Pulling the circuit breaker(s) controlling the power to the autopilot.

For those aircraft with the stall warning wired directly to the autopilot, the autopilot will also disconnect if the stall-warning alarm is present in the aircraft. In most cases, the autopilot disconnect will be accompanied by a 16-beep disconnect aural alert. This tone can be muted by pressing the AP Disconnect switch on the control yoke.

Flight director versus autopilot

The status of the reference bugs, autopilot annunciators, autopilot control head and flight-director steering command bars indicate when the PFD is coupled with the autopilot. A solid magenta heading, altitude, IAS or VS bug indicates that function is currently coupled to an engaged or armed mode of the autopilot or the flight director. A hollow magenta bug indicates that the function is not currently coupled to the autopilot or flight director in an engaged or armed mode. In other words, the autopilot and flight director are ignoring any hollow magenta bug. The flight-director command bars will indicate the required steering of the aircraft to achieve the commanded tracking of the autopilot. In full autopilot mode, both the 'AP' and 'FD' buttons will be lit on the autopilot control panel and 'AP' will be displayed in the autopilot annunciation field on the display, the command bars will be visible and magenta and the aircraft should track those bars very precisely.

In flight-director-only mode, only the 'FD' button (and not the 'AP' button) will be lit on the autopilot control panel and 'FD' will be displayed in the autopilot annunciation field on the display, the command bars will be visible and green, and the pilot will be expected to use the flight controls as required to track those bars. In flight-director-only mode, the pilot is hand-flying the

airplane and is expected to guide the aircraft such that the yellow aircraft reference symbol is aligned with the steering command bars.

The flight-director command bars in a DFC90 autopilot are designed for easy use and improved performance during uncoupled autopilot operations. During coupled operations (both 'AP' and 'FD' buttons lit), pressing the 'FD' button will have no effect. Pressing the 'AP' button in this state will toggle the 'AP' mode on/off. It is a good way to disconnect the servos but continue to have flight-director command bars present. The recommended way to disengage both the 'AP' and 'FD' modes will be via trim or the AP Disconnect switch on the control yoke, as described earlier.

PFD annunciations

The top strip of the PFD is dedicated for autopilot-mode annunciators. Active modes are depicted in green and armed modes are depicted in cyan. Alerts are depicted in yellow and are listed in order of priority. If multiple alerts are received, then the highest priority message is displayed.

Whenever 'UNDERSPEED' or 'OVERSPEED' are displayed while the autopilot is coupled, all engaged (green) autopilot-mode annunciators will flash. Table 8.1 gives a listing of all annunciations that are possible with the DFC90 system.

Table 8.1 DFC90 system annunciations

Normal Annunciations	AP/FD Mode	Lateral		Vertical		Alerts
		Armed	Active	Armed	Active	
	AP (G)	NAV (B)	ROLL (G)	PITCH (G)	ALT (B)	TRIMMING UP (Y)
	FD (G)	NAV APPR (B)	HDG (G)	ALT (G)	GS (B)	TRIMMING DN (Y)
	AP READY (G)	GPSS (B)	NAV (G)	IAS (G)		NAV INVALID (Y)
	CWS (W)	GPSS APP (B)	NAV APPR (G)	VS (G)		GS INVALID (Y)
			GPSS (G)	GS (G)		TC FAIL (Y)

Table 8.1 continued

Normal Annunciations	AP/FD Mode	Lateral		Vertical		Alerts
		Armed	Active	Armed	Active	
		GPSS (Y)				AHRS MISCOMP (Y)
			GPSS APP (G)			NO PFD COMM (Y)
			GPSS APP (Y)			SERVO (Y)
			45⁰ INT (G)			BANK LIMIT (Y)
						AUDIO FAIL (Y)
						MSR FAIL (Y)
Envelope Protection					OVERSPEED (Y)	
					UNDERSPEED (Y)	
Straight & Level			STRAIGHT & LEVEL (G)			
Autopilot Inoperative	AUTOPILOT INOP (Y)					
	AUTOPILOT INOP AHRS FAIL (Y)					
	AUTOPILOT INOP SELF TEST FAIL (Y)					
	AUTOPILOT INOP AHRS ALIGNING (Y)					
	AUTOPILOT INOP TURN COORDINATOR FAIL (Y)					
	AUTOPILOT INOP AHRS MISCOMPARE (Y)					
	NO COMMUNICATION WITH AUTOPILOT (Y)					
Disconnect	AUTOPILOT DISCONNECTED (Y)					

(G) = Green, (W) = White, (B) = Blue, (Y) = Yellow

MULTIPLE-CHOICE QUESTIONS

1. Manometric inputs to the autopilot are typically derived from the:

   (a) Rate gyro
   (b) Centralized air-data computer (CADC)
   (c) Inertial reference system

2. The glide slope would normally be captured from which pitch mode?

   (a) Heading hold
   (b) Vertical speed from above the GS
   (c) Altitude hold from below the GS

3. The DC polarities of a Dutch roll filter's outputs are greatest when the rate of turn is:

   (a) Maximum
   (b) Minimum
   (c) Zero

4. Autopilots incorporating a pitch-trim control allow the pilot to:

   (a) Manually control the desired aircraft attitude
   (b) Manually control the desired aircraft altitude
   (c) Manually control the desired aircraft heading

5. Capturing a VOR radial would normally being achieved from:

   (a) Altitude-hold mode
   (b) Heading-select mode
   (c) Localizer mode

6. Dutch roll occurs with a combination of:

   (a) Yaw and roll disturbance
   (b) Yaw and pitch disturbance
   (c) Roll and pitch disturbance

7. GS deviation signal gain is reduced to account for:

   (a) Narrowing of the localizer beam
   (b) Decreasing airspeed
   (c) Narrowing of the GS beam

8. Yaw dampers are designed to:

   (a) Allow the Dutch-roll frequency to reduce rudder demand
   (b) Allow the Dutch-roll frequency to demand rudder in opposition to the disturbance
   (c) Prevent the Dutch-roll frequency from demanding rudder in opposition to the disturbance

9. Heading-select mode is incompatible with which other roll mode?

   (a) Localizer on course
   (b) Localizer capture
   (c) Glide-slope capture

10. Glide-slope capture is compatible with which other modes?

   (a) VOR capture
   (b) Heading select
   (c) Altitude hold or vertical-speed mode

9 Rotorcraft aerodynamics

This chapter serves as an introduction to aerodynamics and theory of flight for rotorcraft to underpin the study of autopilots and flight guidance systems. The term 'rotorcraft' is used in this book in preference to 'helicopter' or 'rotary wing aircraft' to be consistent with EASA terminology. 'Rotorcraft' in EASA terminology means a heavier-than-air aircraft that depends principally for its support in flight on the lift generated by one or more rotors.

The first sections of this chapter introduce basic rotorcraft features and terminology covering aerodynamics, controls and stability. These topics are described for background information and at a basic level to enable the reader to appreciate the basic principles of rotorcraft aerodynamics. The remaining sections describe rotorcraft automatic control principles and give examples of typical stability-augmentation systems and autopilot systems.

The rotorcraft is a versatile machine that can take off and land vertically, change lateral direction very quickly and hover over a fixed position on any heading. This chapter will address conventional, single main/tail rotor machines only. The study of other rotorcraft configurations, e.g. tandem, 'no tail rotor' and coaxial rotorcraft, is beyond the scope of this book and will be covered by other titles in this book series.

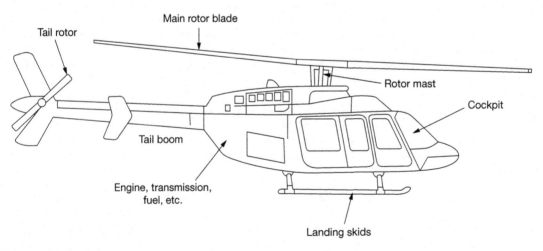

9.1 Rotorcraft main features

ROTORCRAFT FEATURES

The primary features of a rotorcraft are shown in Figure 9.1. The main rotor is used for lift, thrust and lateral control in four directions: forward, aft, left and right. The tail rotor and boom are used for directional control. Rotorcraft engines can be either gas turbines (single/dual or triple) or piston. The engine(s) are normally located above or behind the passenger cabin.

Main rotor

A rotorcraft's main rotors can have two or more blades, depending on its size and role. The drive shaft forms part of the rotorcraft mast. The rotor blades are essentially rotating wings; when rotating, they are referred to as a rotor disc. The landing-gear arrangement can be via skids (with or without flotation devices) or wheels (either fixed or retractable). The main rotor head assembly is complex items of

9.2 Main rotor head

machinery, performing many different functions simultaneously – see Figure 9.2.

Each of the main rotor blades can have its angle of attack varied via control inputs from the pilot (see Figure 9.3). Each blade's angle of attack (AoA) can be varied:

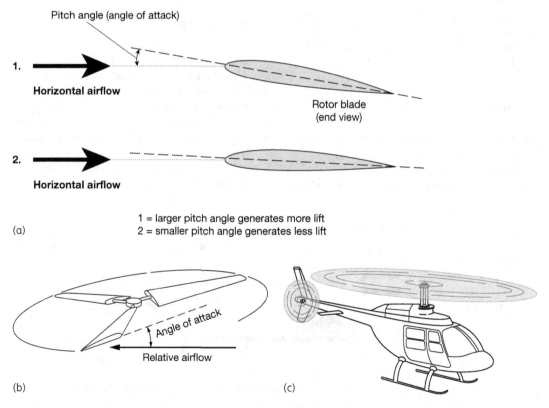

(a)

1 = larger pitch angle generates more lift
2 = smaller pitch angle generates less lift

(b)   (c)

9.3 Rotor blade: (a) Angle of attack (AoA) (b) Lift with AoA (c) Lift with RPM

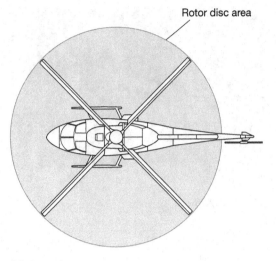

9.4 Rotor disc

- At the same time as the other blades, i.e. collectively
- Per revolution of the blade's position in the cycle

The main rotor head can cause the rotor disc to be pivoted in two planes. This causes the disc plane to be offset or tilted, allowing the rotorcraft to be flown in different directions – see Figure 9.4.

Tail rotor

The tail rotor's functions are for (i) opposing the torque created by the main rotor – see Figure 9.5 – and (ii) yaw/ directional control of the rotorcraft. This is achieved by varying the pitch of the rotor blades.

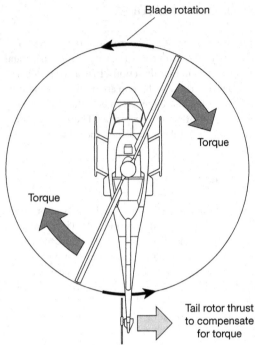

9.5 Main rotor torque

Rotor thrust acts on the tail boom serving as a lever, pivoted on the rotor mast. The tail rotor is driven from the main engine via a main gearbox and tail driveshaft mechanism – see Figure 9.6. Tail rotor blade pitch control is illustrated in Figure 9.7(a); by moving in or out, the spider controls the pitch angle of the blades.

Some rotorcraft have an arrangement known as a Fenestron® (or fantail, sometimes called 'fan-in-fin'), a trademark of Eurocopter. This is a protected, or

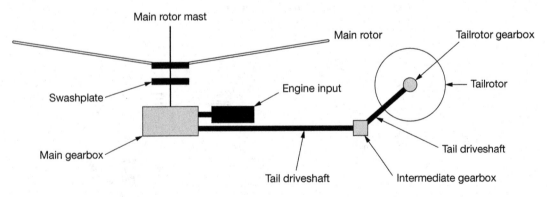

9.6 Transmission overview

9.7 Tail rotor: (a) Conventional (b) Fenestron®

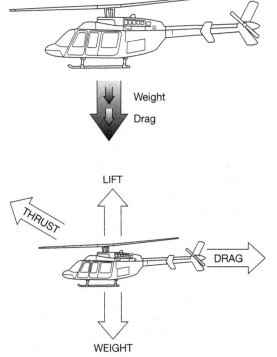

9.8 Primary forces: (a) Hover (b) Forward fight

shrouded, tail rotor that operates much like a ducted fan, with several advantages compared with the conventional tail rotor:

- Reducing tip vortices, and associated noise
- Protecting the tail rotor from damage
- Protecting ground crews from the hazard of a spinning rotor

While conventional tail rotors typically have two or four blades, Fenestrons have up to 18 blades. These sometimes have variable angular spacing, so that the noise is distributed over different frequencies. The shroud allows a higher rotational speed than a conventional rotor, resulting in smaller blades.

To maintain the rotorcraft in a hover, and at constant altitude, lift and thrust must be equal to the rotorcraft weight and blade drag, as in Figure 9.8a. Whilst in flight, the forces acting on a rotorcraft are: weight, drag, lift and thrust – see Figure 9.8b. The latter two are created from the main rotor disc; the rotor head's mechanical features are used for varying the pitch of each blade and tilting of the rotor disc.

KEY POINT

Each rotor blade creates its own lift and drag.

KEY POINT

Main-rotor torque is countered by tail-rotor thrust.

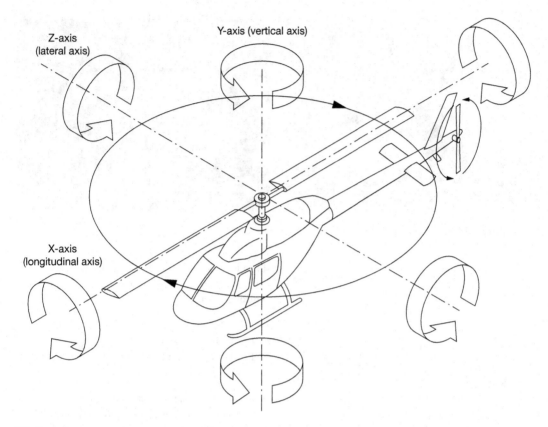

9.9 Control axes

The pilot has control of the rotorcraft in three axes: vertical, lateral and longitudinal – see Figure 9.9. The rotorcraft is controlled about each axis as follows:

- X-axis, main rotor tilt
- Y-axis, tail rotor thrust
- Z-axis, main rotor tilt

PRIMARY FLYING CONTROLS

The primary flying controls for the rotorcraft are the collective lever, cyclic stick and tail rotor pedals – see Figure 9.10. The collective lever is used to change the pitch/AoA of each main rotor blade by an equal amount at the same time; this allows the pilot to change the rotorcraft's lift and thrust. The cyclic stick is used to change the pitch/AoA of each main rotor blade by varying amounts within specific cycles of the blade's position. This allows the pilot to control

the rotorcraft around the X and Z axes. The tail rotor pedals are used to change the pitch/AoA of the tail rotor blades by an equal amount at the same time, to change the thrust from the rotor; this allows the pilot to control the rotorcraft around the Y-axis.

Main rotor blades

Lift and/or thrust control of the rotorcraft is achieved by varying the pitch angle of each of the main rotor blades collectively; this can be achieved by a swash plate or spider system. The swash-plate system incorporates an upper and lower section, as seen in Figure 9.11, with a bearing between each section. The lower section can tilt in any direction, but does not rotate; the upper section tilts to match the lower section's position, and rotates at the same speed as the rotor. The lower section is moved by the pilot's control inputs via control rods. The rotor disc is tilted in

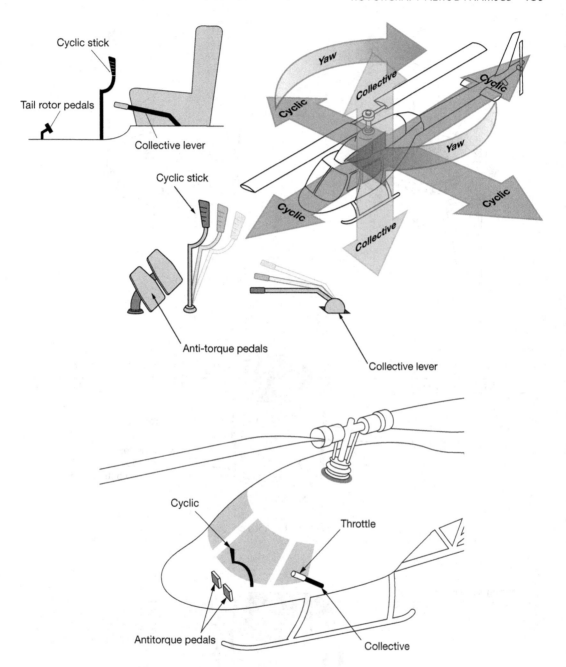

9.10 Cockpit controls

the required direction in response to pilot control inputs into the lower section of the swash plate. The up/down movement of the upper section rods is translated into each blade's pitch angle. In the spider system – Figure 9.12 – the leading edges of the blades are connected to the arms of a 'spider'. The blades are connected to a cylinder or universal joint arrangement; the blades are moved individually by tilting the spider or in unison by raising/lowering the spider.

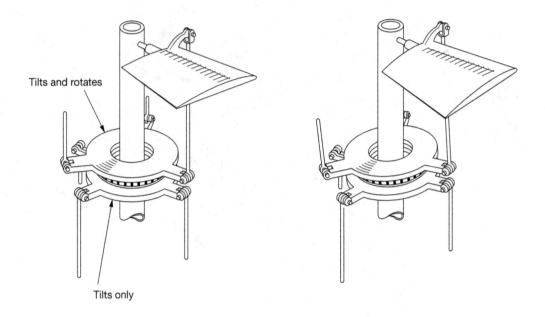

Tilts and rotates

Tilts only

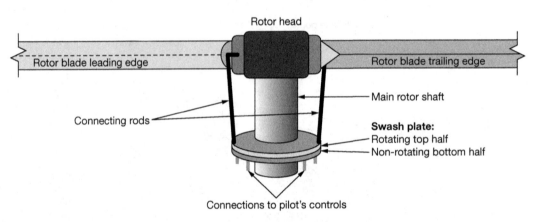

Rotor head

Rotor blade leading edge

Rotor blade trailing edge

Main rotor shaft

Swash plate:
Rotating top half
Non-rotating bottom half

Connecting rods

Connections to pilot's controls

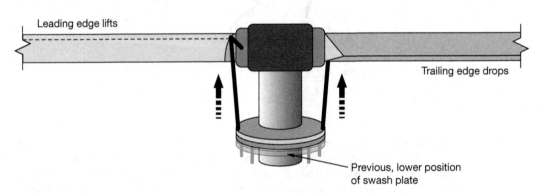

Leading edge lifts

Trailing edge drops

Previous, lower position
of swash plate

9.11 Swash plate principles

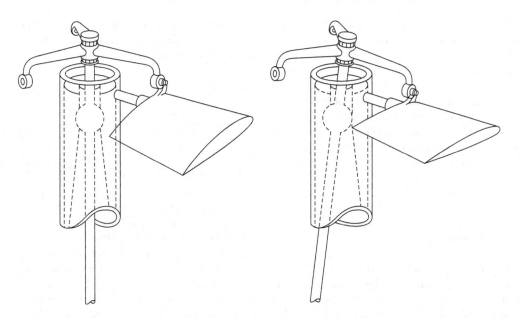

9.12 Spider principles

Collective control

Moving each blade by the same amount allows the pilot to vary the amount of lift generated by the entire disc. Lifting the collective control will increase blade pitch equally on every blade, as seen in Figure 9.13.

The collective lever also incorporates the throttle control; although most rotorcraft engines are maintained at constant RPM, this control input is used to adjust torque from the main rotor. Rotorcraft rotors are designed to operate in a narrow range of RPM. Depending on the rotorcraft model, the engine speed

9.13 Collective linkage and control

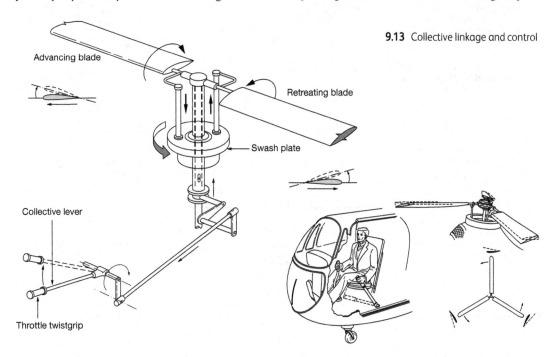

Advancing blade

Retreating blade

Swash plate

Collective lever

Throttle twistgrip

will vary. On a large rotorcraft the engine RPM is typically 20,000 RPM, producing approximately 2,000 shaft horse power (SHP) per engine. The engine's output is applied to the rotor via a gearbox, producing a rotor speed of approximately 250 RPM (about 4 times a second) at the main rotor hub, without loss of torque or power. Depending on the rotor blade length, the tip speed will be in the order of 350 miles per hour; the longer the blade, the faster the tip speed. The blade is designed to not produce lift at the tip; most of the lift is produced in the centre section of the blade.

Cyclic control

The cyclic control is used to tilt the rotor disc, and move the rotorcraft in lateral directions, as in Figure 9.14. The cyclic control modulates each blade's angle of attack (AoA) as it moves through the air. A higher angle of attack increases lift for a given relative airspeed, a lower angle of attack decreases lift.

Whereas the collective changes the angle of attack of all of the blades at the same time, i.e. 'collectively',

thus changing the overall lift/thrust from the rotor disc, the cyclic control modulates each of the blades as they rotate in the 'cycle' – see Figure 9.15.

By increasing a blade's AoA as it moves towards one point in the rotor disc, and decreasing the AoA as it moves to the opposite side, lift on one side is of the disc is increased. Moving the cyclic control forward, the blades will have a higher AoA as they approach the rear of the disc and a reduced AoA as they approach the front of the disc. This results in more lift in the back of the rotor disc, the disc tilts (or pitches) forward, and the rotorcraft travels in a forward direction. The same principle applies to controlling the rotorcraft for left/right directional control.

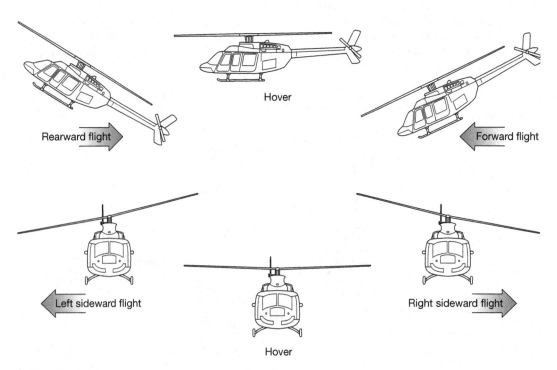

9.14 Cyclic control

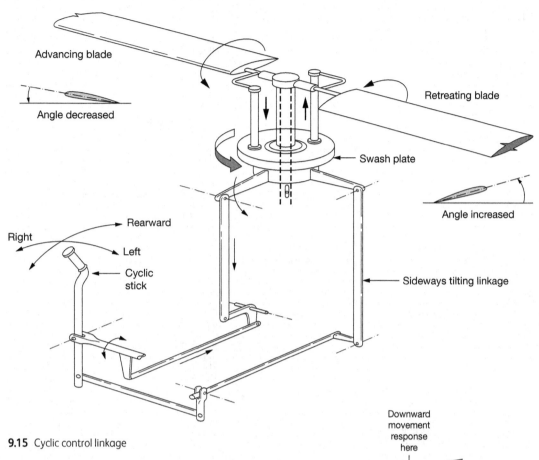

Advancing blade

Angle decreased

Retreating blade

Swash plate

Angle increased

Rearward

Right

Left

Cyclic
stick

Sideways tilting linkage

9.15 Cyclic control linkage

AERODYNAMICS

Gyroscopic effects

When considering the main and tail rotor blades as discs, all laws of the gyroscope apply as previously described earlier in this book – see Figure 9.16. When a control input is made into the rotor disc in the form of a downward force at point A, the effect of precession will occur at 90 degrees to the control input, and the response will be at point B. This will require a corresponding control input from the pilot.

Dissymmetry of lift

With the rotorcraft hovering in zero wind conditions, the relative airflow is the same for each main rotor blade. When the rotorcraft is travelling in a lateral

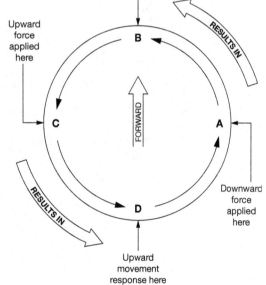

Downward
movement
response
here

Upward
force
applied
here

B

RESULTS IN

C

FORWARD

A

Downward
force
applied
here

RESULTS IN

D

Upward
movement
response here

9.16 Gyroscopic effects

9.17 Dissymmetry of lift

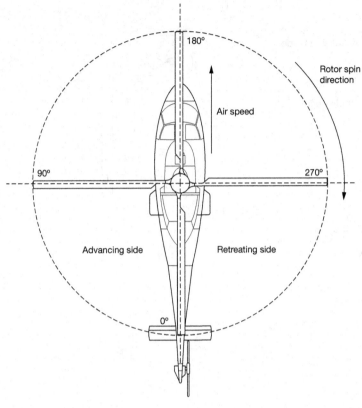

9.17 Dissymmetry of lift

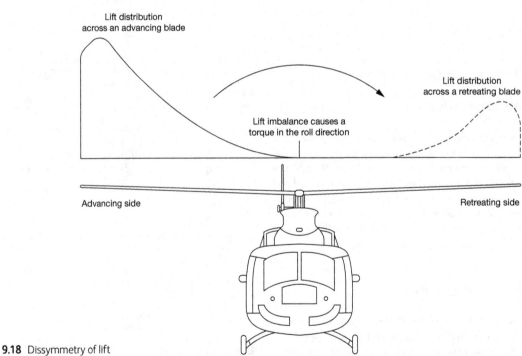

9.18 Dissymmetry of lift

direction, e.g. forwards, the advancing half of the rotor disk and the retreating half give rise to dissymmetry of lift – see Figures 9.17, 9.18 and 9.19. This phenomenon is caused by relative airflow over the blades being added to the rotational relative airflow on the advancing blade, and subtracted on the retreating blade. This effect will also occur in the hover when the rotorcraft is operating with any wind conditions. To equalize the lift of each blade, they are free to 'flap' via hinges to change the AoA on a cyclic basis.

Blade flapping

The advancing blade is exposed to airflow from two sources: forward flight velocity and rotational airspeed (of the rotor); the blade responds to the increase of speed by producing more lift. The blade flaps (or climbs) upward, and the change in relative airflow and

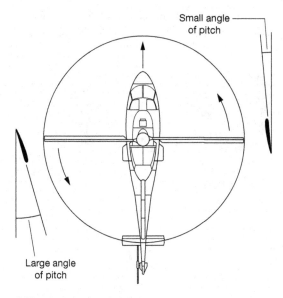

9.19 Dissymmetry of lift

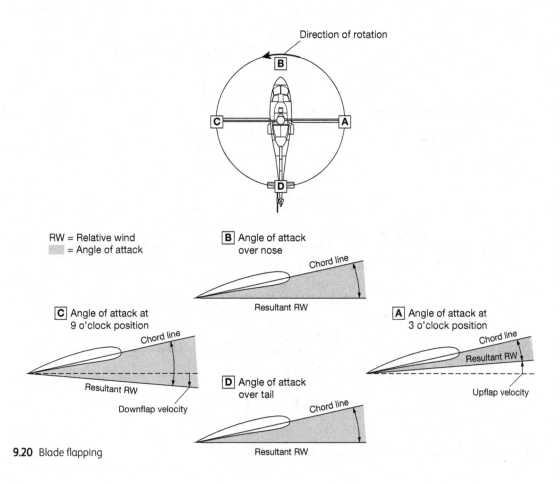

9.20 Blade flapping

angle of attack reduces the amount of lift that would have been generated. The resulting larger angle of attack retains the lift that would have been lost because of the reduced airspeed. In the case of the retreating blade, the opposite is true. As it loses airspeed, reducing lift causes it to flap down (or settle), thus changing its relative airflow and angle of attack (see Figure 9.20).

Since the tail rotor also has advancing and retreating blades during forward flight it will also experience dissymmetry of lift. This is corrected for by hinging the blades such that they 'flap'. Blade flapping is the upward/downward movement of a rotor blade, which, in conjunction with cyclic feathering, causes dissymmetry of lift to be eliminated.

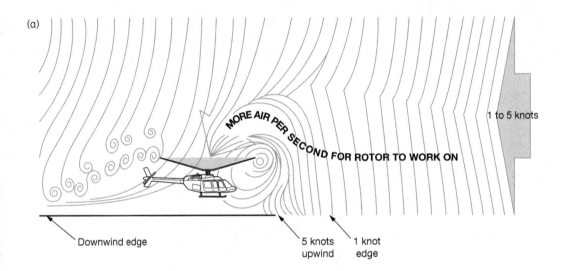

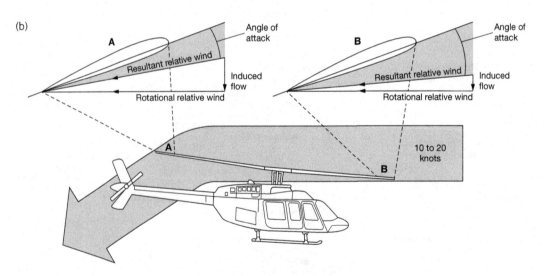

9.21 Translational lift: (a) Low airflow conditions (b) Higher airflow conditions

Translational lift

Translational lift is present with any horizontal flow of air across the rotor disc. This increased flow is more pronounced when the relative airspeed reaches approximately 10 to 20 knots (see Figure 9.21).

A rotorcraft in forward flight, or hovering in a headwind or crosswind, has a greater amount of airflow entering the aft portion of the main rotor blade. The angle of attack is therefore less, and the induced airflow is greater at the rear of the rotor disc. As the rotorcraft accelerates through this speed, the rotor disc moves out of its own vortices and into relatively undisturbed air. The relative airflow is now becoming more horizontal; this reduces induced flow and drag, with a corresponding increase in angle of attack and lift. The additional lift created at this speed is the 'effective translational lift' (ETL).

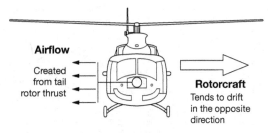

9.22 Translating tendency

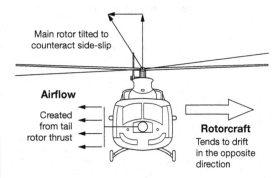

9.23 Translating tendency

> **KEY POINT**
>
> Effective transitional lift is recognizable in flight by a transient induced vibration and increased rotorcraft performance.

When the rotorcraft flies through translational lift, the relative air flowing through the main rotor disc and over the tail rotor disc becomes less turbulent and hence aerodynamically efficient. As the tail rotor disc's efficiency increases, more thrust is produced, causing the aircraft to yaw. If no corrections are made by the pilot (depending on the rotational direction of the rotor) the rotorcraft pitches up, and rolls to the right. This is caused by the combined effects of dissymmetry of lift and transverse flow. Translational lift can also be achieved in the hover when the wind speed is approximately 15 to 25 knots.

A rotorcraft that is hovering in a crosswind, or headwind, or in forward flight has an increased amount of airflow entering the aft section of the rotor disc. The AoA is therefore reduced and induced flow increased at the rear of the disc.

Translating tendency

During hover, the rotorcraft has a tendency to drift laterally due to the thrust of the tail rotor – see Figure 9.22. The pilot can compensate for this via the cyclic control, i.e. by tilting the main rotor disc. This results in the main rotor force compensating for the tail rotor thrust.

Rotorcraft type design usually includes one or more features which help the pilot compensate for translating tendency. Flight control rigging may be designed so that the rotor disc is offset when the cyclic control is centred (see Figure 9.23). Alternatively, the collective pitch control system may be designed so that the rotor disc tilts slightly left as collective pitch is increased to hover the rotorcraft. Finally, the main transmission arrangement may be mounted so that the drive shaft (or mast) is tilted slightly when the fuselage is laterally level.

Autorotation

Autorotation facilitates a controlled descent through to an emergency landing in case of power-plant

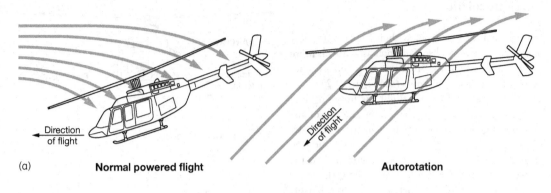

(a) **Normal powered flight**   **Autorotation**

9.24 Autorotation: (a) Normal forward flight and autorotation (b) Airflow during autorotation

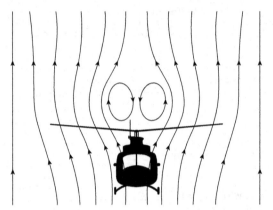

(b)

failure. During autorotation – see Figure 9.24(a) – the relative upward airflow causes the main rotor blades to rotate at their normal speed; the blades are 'gliding' in their rotational plane. The airflow pattern in autorotation is shown in Figure 9.24(b).

> **KEY POINT**
>
> Autorotation means a rotorcraft flight condition in which the lifting rotor is driven entirely by action of the air when the rotorcraft is in motion.

> **KEY POINT**
>
> During autorotation, the upward flow of relative airflow allows the main rotor blades to rotate at their normal speed, i.e. the blades are effectively gliding in the disc's plane of rotation.

Hovering

This is a rotorcraft feature that occurs at any time the relative speed between the rotorcraft and ground surface is zero. Referring to Figure 9.25, this can mean:

- Zero ground speed hover
- Zero airspeed hover
- Hovering alongside a moving ship
- Zero surface water speed hover

Ground effect

This is encountered when operating near the ground (within half rotor diameter), as in Figure 9.26. Improved rotor disc performance is achieved due to the interaction of the surface with the airflow from the rotor system. This effect is more pronounced when approaching the ground.

When hovering away from any ground effect, large rotor tip vortices are generated and the rotor disc

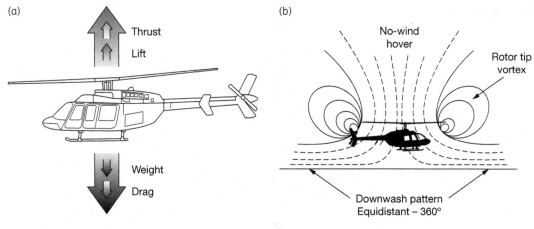

9.25 Hovering (a) Forces (b) Vortices

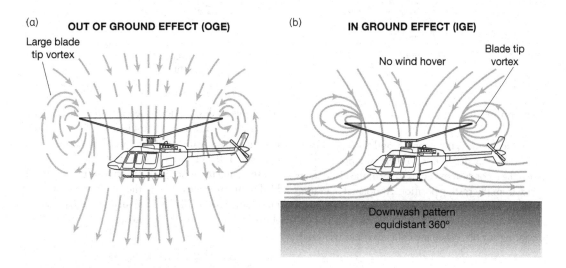

9.26 Ground effect: (a) Out of ground effect (OGE) (b) In ground effect (IGE)

downwash is nearly vertical. When approaching the ground, these vortices are reduced as a consequence of the change in downwash direction to a more horizontal direction. The varying effect of proximity to the ground means that the pilot has to compensate for variations in thrust as a function of height above the ground.

Pendulum effect

The fuselage of a rotorcraft can be considered as being suspended from a single point, as seen in Figure 9.27. Since it has mass, it is free to move either longitudinally or laterally in the same way as a pendulum. This pendulum effect can be overstated by the pilot over controlling the cyclic control inputs in the hover and forward/rearward flight.

9.27 Pendulum effect

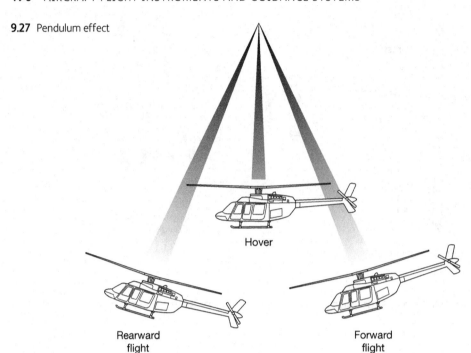

Hover

Rearward
flight

Forward
flight

Disc coning

When the engine is not running, and the rotor blades are at rest, they droop due to their weight and span. When the rotor system begins to turn, the blade starts to rise from the resting position. At normal operating RPM, the blades extend straight out due to centrifugal force; at this point they are at a neutral AoA and not producing any lift. When the collective lever is raised, the blades all develop increased lift, and the disc rises into a coned position – see Figure 9.28. Increased coning causes a decrease of total lift because of a decrease in the effective disc area.

Referring to Figure 9.29, air arrives at the forward blades from below, and the AoA subsequently increases. The aft blades have airflow arriving from above, thereby reducing the AoA.

> **KEY POINT**
>
> There is a lift imbalance between forward and aft blades in forward flight due to coning.

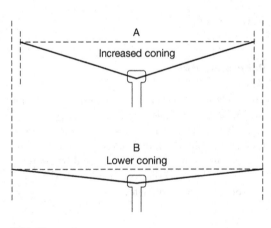

9.28 Disc coning

Coriolis Effect

When a rotor disc moves upward, thereby increasing the cone angle, e.g. due to increasing speed and lift, the centre of mass of the blade moves closer to the axis of rotation, and the rotor velocity increases. Conversely, when the blade moves downward, its centre of mass moves further from the axis of rotation

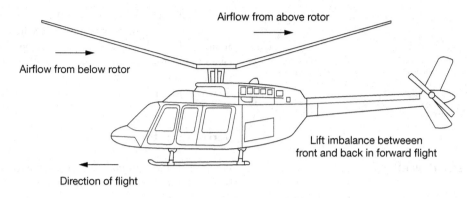

9.29 Disc coning

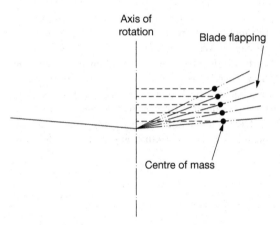

9.30 Coriolis Effect

Induced flow

As the rotorcraft accelerates in forward flight, induced flow decreases to near zero at the forward disc area and increases at the aft disc area. This increases the angle of attack at the front disc area, causing the rotor blade to flap up, and reduces angle of attack at the aft disc area, causing the rotor blade to flap down (see Figure 9.31). Because the rotor acts like a gyro, maximum displacement occurs at 90 degrees to the direction of rotation. The result is a tendency for the helicopter to roll slightly as it accelerates through approximately 20 knots or if the headwind is approximately 20 knots.

and the rotor velocity decreases. This is termed the Coriolis Effect – see Figure 9.30. Depending on the design of the rotor system, the acceleration/deceleration actions of the rotor blades are absorbed by either dampers or the blade structure itself.

> **KEY POINT**
>
> Coriolis Effect is the change in blade velocity to compensate for the change in distance of the centre of mass from the axis of rotation as the blades flap.

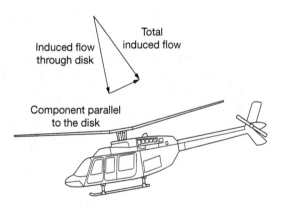

9.31 Induced flow

Transverse flow effect

An imbalance of lift is caused by the flow of air through the rotor while the rotor is tilted into the airflow during take-off. The air underneath the rotor disc is recirculated due to ground effect, resulting in a higher angle of attack. Conversely, air on the upper side of the disc passes through and is accelerated clear of the disc, resulting in a lower angle of attack. This imbalance of lift typically occurs at between 12 and 15 knots of relative airflow, i.e. from forward flight or crosswinds in a hover.

Transverse flow effect will be noticed by a vibration of the main rotor caused by a lift imbalance between the left and right sides of the rotor disc. Transverse flow increases vibrations of the helicopter at airspeeds just below effective translational lift on take-off and after passing through effective translational lift during landing. To counteract transverse flow, a cyclic input needs to be made.

Vortex ring state

In a normal out-of-ground-effect hover, the rotorcraft is able to remain in a stable condition by forcing air down through the main rotor. Some of this air is recirculated at the rotor tips; a phenomenon common to all aerofoils known as tip vortices. These relatively small vortices consume engine power without any productive lift, resulting in a small loss of rotor efficiency. When the helicopter begins to descend vertically, it settles into its own downwash, and the tip vortices increase. An increasing amount of power developed by the engine is now wasted in accelerating the air around the rotor, as seen in Figure 9.32.

During descent, air vortices can form around the main rotor, causing a situation known as vortex ring

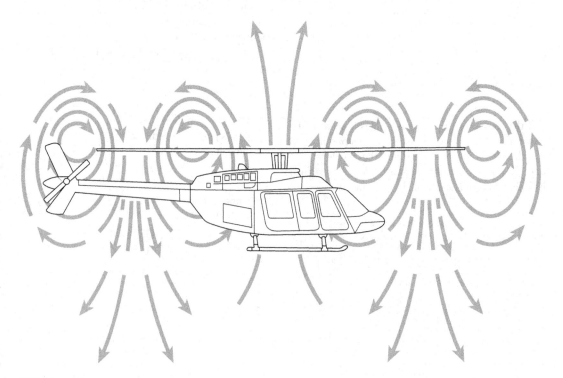

9.32 Vortex ring state

state (VRS); also referred to as 'settling with power'. In this situation, the airflow downwash through the main rotor is subjected to the following sequence of directional changes:

- Outwards
- Inwards
- Down through the rotor again

This recirculation of airflow can counter much of the rotor's lifting force, causing a significant loss of altitude. Applying more power and increasing collective pitch accelerates the downwash through which the main rotor is descending, compounding the situation.

Furthermore, the rotorcraft may descend at a rate exceeding the normal downward induced-flow rate of the rotor's inner blade sections. Consequently, the airflow at the inner blade sections is upward relative to the rotor disc. This produces a secondary vortex ring in addition to the normal tip-vortices. This secondary vortex ring is formed about the point on the blade where the airflow changes from up to down. The result is unstable, turbulent airflow over a large area of the disc; rotor efficiency is reduced even though power is still being supplied from the engine, often leading to hazardous conditions.

MULTIPLE-CHOICE QUESTIONS

1. Translating tendency can be compensated for by:

   (a) Increasing thrust from the tail rotor
   (b) Tilting the main rotor to counteract the side-slip
   (c) Increasing torque from the main rotor

2. The collective lever is used primarily for:

   (a) Lateral control
   (b) Yaw control
   (c) Thrust and lift

3. Main rotor torque is countered by:

   (a) Autorotation
   (b) Coriolis Effect
   (c) Tail rotor thrust

4. During hover, the rotorcraft has a tendency to drift laterally due to the thrust of the:

   (a) Main rotor
   (b) Engines
   (c) Tail rotor

5. Coriolis Effect occurs when the centre of mass of the blade moves:

   (a) Away from the axis of rotation, and the rotor velocity increases
   (b) Closer to the axis of rotation, and the rotor velocity increases
   (c) Closer to the axis of rotation, and the rotor velocity decreases

6. When the rotorcraft flies through translational lift, the relative air flowing through the main rotor disc and over the tail rotor disc becomes:

   (a) Less turbulent and hence aerodynamically efficient
   (b) More turbulent and hence aerodynamically efficient
   (c) Less turbulent and hence aerodynamically inefficient

7. Dissymmetry of lift is caused by relative airflow over the blades being:

   (a) Added to the rotational relative airflow on the advancing and retreating blades
   (b) Less turbulent and hence aerodynamically efficient
   (c) Added to the rotational relative airflow on the advancing blade, and subtracted on the retreating blade

8. The tail rotor's functions are for opposing the:

   (a) Torque created by the main rotor and vertical control of the rotorcraft
   (b) Torque created by the main rotor and directional control of the rotorcraft
   (c) Lift created by the main rotor and directional control of the rotorcraft

9. The cyclic control changes the:

    (a) Main rotor's thrust direction
    (b) Tail rotor's thrust direction
    (c) Lift and/or thrust control of the rotorcraft

10. As the rotorcraft accelerates in forward flight, induced flow:

    (a) Increases to near zero at the forward disc area and increases at the aft disc area
    (b) Decreases to near zero at the forward disc area and increases at the aft disc area
    (c) Decreases to near zero at the aft disc area and increases at the forward disc area

10 Rotorcraft automatic flight control

Automatic flight-control systems (AFCS) for rotorcraft include the various technologies designed to control and/or improve basic stability, handling qualities and lateral/vertical guidance. Furthermore, the AFCS can be designed and utilized for specific missions and/or manoeuvres, allowing these to be flown automatically. In general terms, the AFCS allows the pilot to become the 'cockpit resource manager'. There are various configurations of AFCS; some or all of the following features may be incorporated in various ways into the rotorcraft's design, depending on the type of design and rotorcraft operation.

STABILITY

The static stability of a rotorcraft is a measure of its tendency to return to its original flight path after a displacement or disturbance (see Figure 10.1). If the disturbance causes the nose to be displaced upwards, and the rotorcraft has a tendency to return to its initial attitude, it has static stability. For the same nose-up disturbance, if the rotorcraft has a tendency to deviate even more from its initial attitude, it is statically unstable.

Dynamic stability describes the rotorcraft's ongoing reaction once the disturbing force is removed (see Figure 10.2). This ongoing reaction is in the form of oscillations; the reaction can be in one of three forms:

- Dynamic instability: the rotorcraft continues to move in the direction of the force with increasing deviations.

- Neutral stability: the rotorcraft continues to move in the direction of the force with equal deviations.
- Dynamic stability: the rotorcraft returns to the equilibrium position, with decreasing deviations.

ATTITUDE-HOLD SYSTEM (AHS)

This system will maintain the commanded attitude in each of the three axes, i.e. pitch, roll and yaw. The AHS normally requires control-position information to permit manoeuvring. Some attitude-hold systems may revert to control-rate damping during manoeuvres commanded by the pilot.

Stability-augmentation system (SAS)

This system normally uses rate gyros to dampen rotorcraft oscillations; it can also include integration of rate-gyro output signals to provide short-term pseudo attitude hold. The stability-augmentation system is generally used to dampen changes in attitude without regard to pilot demand for manoeuvring. The system is intended to improve the basic flying qualities of the rotorcraft; it should not affect static stability.

Automatic stabilization equipment (ASE)

ASE uses a vertical gyro or other attitude reference input to provide a long-term attitude-hold function. Rate damping may also be incorporated to account for

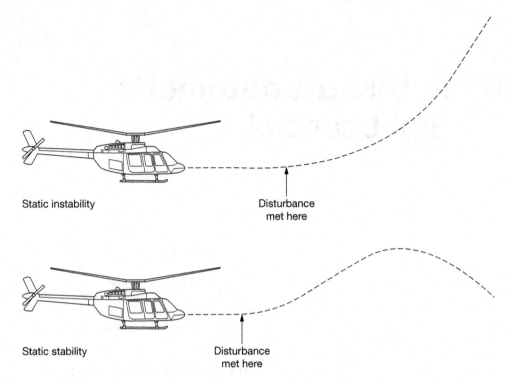

Static instability

Disturbance met here

Static stability

Disturbance met here

10.1 Static stability

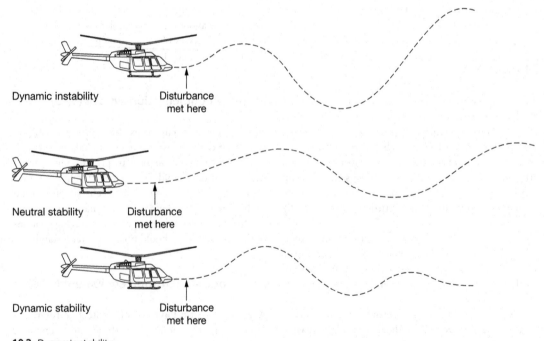

Dynamic instability

Disturbance met here

Neutral stability

Disturbance met here

Dynamic stability

Disturbance met here

10.2 Dynamic stability

the pilot making a control input. Other ASE features include heading and altitude hold using the compass and air data systems as reference sensors.

AUTOPILOTS

The rotorcraft autopilot, or automatic flight control system (AFCS), can include numerous functions in addition to those previously described:

- Airspeed hold
- Automatic transition to a hover
- Capture and tracking of a lateral course (e.g. VOR)
- Coupled approaches (e.g. GPS, ILS)

The AFCS can be defined as the various flight controls that are moved automatically by a device other than the pilot. The amount of the control movement that the AFCS can influence is referred to as its 'authority'. This is normally given as percentage of total control movement in each axis, or an equivalent deflection of a given flying control.

The inner loop of the system is used to make small flying-control adjustments to counter internal/external disturbances, e.g. changes to centre of gravity, wind gusts and so on. The outer loop of the system is used to make additional flying-control adjustments (superimposed on the inner loop adjustments) to enable particular conditions to be maintained or changed, e.g. heading hold, altitude select and so on. Figure 10.3 shows the typical relationships of inner and outer loop control.

Some AFCS include a 'fly-through' feature – this enables the pilot to make a change to the rotorcraft's flight path without having to disengage and then re-engage the AFCS. The fly-through feature makes control changes from the AFCS invisible to the pilot when manoeuvring; this is sometimes referred to as system 'transparency'. The ability of the AFCS to

perform a specific function using various sensors is called coupling, i.e. a coupled ILS approach will require certain AFCS lateral and vertical functions, as well as usable GPS and/or ILS signals.

There are a variety of autopilot modes for outer-loop control, ranging from simple two-axis control with analogue sensors and systems through full-authority digital AFCS. Ideally these modes should operate in conjunction with a flight director, so that the pilot can monitor the correct operation of the AFCS, and also the mode to be flown manually with visual guidance.

TEST YOUR UNDERSTANDING 10.1

Explain the terms fly-through, inner loop and outer loop.

SERVO CONTROL

The individual axis (also known as a lane or channel) of the AFCS, i.e. pitch, roll or yaw, includes all the elements of that part of the AFCS. Each channel of the AFCS will cause a specific flying control to be moved by an actuator (servo motor) connected to the flight controls. The number and location of these servos depends on the type of system installed. Actuator trim is needed to adjust the flying-control position, either to a predetermined position prior to a specific manoeuvre or to maintain it near its centre of travel. Trim may be achieved automatically or manually.

The method by which the AFCS controls the rotorcraft is through two types of actuators. The first type moves in series with the normal flight controls and is normally of limited authority, with a high rate of actuator movement. Parallel actuators are generally

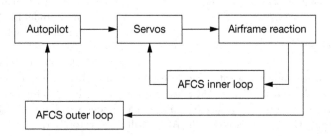

10.3 Inner and outer loop

slow moving/low rate of actuator movement devices with high authority. Movement of an actuator is sensed by a transducer, typically a linear or rotary variable displacement transducer (LVDT or RVDT). Actuator position is then fed back into the AFCS computer's servo control system.

CONTROL AXES

Although each axis of the rotorcraft can be controlled simultaneously, each axis/channel has unique characteristics that have to be considered as part of the AFCS design.

- Two-axis autopilots control the rotorcraft in pitch and roll
- Three-axis autopilots control the rotorcraft in pitch, roll and yaw
- Four-axis autopilots control the rotorcraft in pitch, roll, yaw and collective.

Pitch channel

The performance of the AFCS will be strongly influenced by the variation of pitch attitude with speed. In such a case an artificial airspeed-hold augmentation or a means of producing positive 'cockpit' static stability by altering the length of control runs may be incorporated. Both cyclic stick force and position are important in determining static stability.

Roll channel

Many automatic stabilization equipment installations feature an angle of bank (AOB) hold function which may operate over a limited range of roll attitude or over the full range. The effect of an AFCS on static lateral stability will depend upon the influence of yaw-pedal movement on bank-angle hold.

Yaw channel

The directional stability of a rotorcraft is arguably the most complex function for an AFCS because the mode of operation must change between hover and forward flight. Generally speaking, the AFCS will provide heading hold in the low speed regime and a turn-coordination/heading-hold facility above certain airspeeds. Some AFCS retain the use of the yaw channel to provide simple heading hold in forward flight while others use turn-coordination in yaw and hold heading by adjusting AOB. In forward flight the disengagement of heading hold or the changeover from heading hold to turn coordination can be a function of:

- Angle of bank
- Lateral cyclic displacement
- Yaw pedal displacement
- Operation of a yaw force link (or micro switch).

DESIGN CONSIDERATIONS

The hover/forward-flight-mode change is normally a function of airspeed and will have been evaluated in a variety of off-trim conditions with various degrees of side-slip during turns, and so on. The initiation of a turn in forward flight can be problematic, especially for systems which achieve turn coordination via side-slip transducers. Some rotorcraft use a feed forward of the rate of lateral cyclic input into the yaw channel to initiate a yaw rate during turn entry. The extra yaw input is then gradually reduced (or 'washed' out) as the steady-state turn coordination takes over again. The operation of such a system will have been evaluated during flight test for slow and rapid entries into turns at various bank angles since imperfect operation can lead to problems with yaw stability and/or control.

For all three AFCS channels, dynamic stability has to be considered during the design and certification programme. The fundamental requirement for an AFCS in terms of dynamic stability is to automatically suppress the nuisance modes of the natural rotorcraft response such as long-term pitch oscillations (termed a 'phugoid') or lateral/directional oscillations (termed 'Dutch roll'). The trim condition may be defined in terms of attitude, heading, airspeed, altitude or any other parameter which may be relevant to the particular AFCS function.

The features and functionality of a given AFCS will have been carefully evaluated and mission-related since they may force the pilot to employ a flying technique which is not ideal for the intended purpose of the rotorcraft.

TEST YOUR UNDERSTANDING 10.2

Explain the terms feed forward and washed out.

MULTIPLE-CHANNEL SYSTEMS

Multiple-channel systems refer to the use of more than one AFCS lane (or channel) for a given axis of control:

- Simplex systems use one lane/channel
- Duplex systems use two lanes/channels
- Triplex systems use three lanes/channels
- Quadruplex systems use four lanes/channels

A simplex system will have at least part of the system relying on only one channel and/or component in the AFCS, e.g. a vertical gyro or an actuator. The term 'simplex' does not include common-use rotorcraft components such as flight-control tubes or swash plates. The duplication of simplex capabilities results in a duplex system, e.g. the AFCS has two independent vertical gyros for redundancy.

In a system with three or more channels, a subsystem will be installed to compare outputs from the various channels, with the intention of disconnecting faulty channels to provide increased AFCS integrity. Some systems have a separate system to act as if it were a control channel, but without actually moving the control surface. This subsystem uses a synthetic model of the AFCS control laws for comparison with an active channel's output to a flying control. This technique is sometimes referred to as 'model following'.

Comparing signals between channels, either between two computers, actuators or other devices, is used within multi-channel AFCS to check if the given output signals are the same (within a given tolerance). This is generally used as a failure-protection technique, i.e. to disconnect a channel when the output is determined to be faulty.

CONTROL LAWS

Overview

The rotorcraft AFCS is based on the servo-control system described earlier in this book. To command the movement of a control surface by a predetermined amount requires a specific control law or 'response command'. For example, in a rate-command system, a 25mm lateral cyclic displacement to the left may always give five degrees per second roll rate, regardless of airspeed. Other AFCS could give a constant stick force/acceleration at speeds above 50 knots and give a constant attitude change per unit deflection below that speed.

Gain is the ratio of output to input of a control system, which normally translates to the rate of movement of an actuator. Gain can be short or long term depending on the control laws; long-term motion or response of the rotorcraft is typically more than five seconds. Airspeed sensors can be employed to modify the response of controls with high authority, e.g. a movable horizontal stabilizer.

The insertion of a control signal in the system somewhere downstream of any other signal is referred to as 'feed forward'. When the movement of a flight control is sensed, it can be fed forward into the AFCS in such a manner as to cause an initial movement of the actuator in the desired direction; this technique is called 'control augmentation'. This is done to offset the sensation of a reduction in control response that would occur due to the AFCS damping the motion when it senses a disturbance. 'Quickening' is a technique typically used for control augmentation. This is normally achieved by inserting a command input directly into an actuator before the computation has a chance to react with a response.

One technique used for faster fault detection, as well as failure protection, is cross-coupled feedback. This involves feeding the position of one actuator into both its own computer and the other computers of a multiplexed system.

TEST YOUR UNDERSTANDING 10.3

Explain the terms gain, quickening and feed forward.

Runaway

Runaway of an AFCS is an undesired event; it can be defined as the movement of an actuator to an undesired position. The runaway (or hard-over) results in the rapid movement of an actuator to the end of its travel, which (given the conditions) would not have been normally commanded by the system. This is generally the result of a failure condition, and is a serious event due to the rapid rate of control surface movement. Automatic retrimming of an actuator is often incorporated in a simplex system to minimize the effects of a runaway.

To allow the pilot to deal with a runaway, the AFCS can normally be disengaged at any time during flight, and on the ground. Disengagement of the AFCS can be required at any time, either manually by the pilot, or automatically in case of a system fault. The pilot's primary disconnect facility is normally a push button/ momentary switch on the cyclic control.

An example of automatic disengagement of the inner loop would be attitude hold when the cyclic is moved; this must not introduce any unwanted control responses. An example of automatic disengagement of an outer-loop control would be heading hold disengaging whenever the compass system is aligned or slewed.

MAINTENANCE CONSIDERATIONS

The maintenance engineer could be faced with a variety of reported AFCS defects; these could be from pilot reports – 'pireps' – and/or as a result of ground testing. The following is not an exhaustive list of AFCS defects, but an overview of potential problems caused by one or more AFCS defects.

> ### KEY POINT
>
> These considerations apply to both aeroplanes and rotorcraft.

Passive failures: The AFCS disconnects and ceases to function without causing any unsafe or intrinsic rotorcraft response. There may not be any noticeable effect on the rotorcraft in the short term. Once engaged, the AFCS should not cause any uncontrolled disturbances to the rotorcraft's stability.

Runaways: An actuator is actively driven to an unwanted position. The extreme case is a full-scale, maximum-rate deflection known as a hard-over. Runaways may occur in more than one axis simultaneously.

Oscillatory failures: These generally result from the failure of a primary sensor, e.g. the vertical gyro, or a component in the feedback loop, e.g. an LVDT. This failure may be so severe that the pilot is only able to disengage the system using the cyclic disconnect switch.

Actuator saturation: The inability of the AFCS to function correctly due to the relevant actuator being at full travel.

Degraded modes: The modes of operation are less than fully operational. For example, with the roll channel failed, it may be still possible for the AFCS to have an altitude control mode, but not a coupled ILS approach.

CASE STUDY – GENERAL AVIATION AUTOPILOT

There are many different systems in operation; these vary depending on the physical size of the rotorcraft, and its intended operational use. Cool City Avionics, based in Mineral Wells, Texas, USA has developed a range of automatic flight control and stability-augmentation systems. The SAS-100 is a stability-augmentation system designed to improve basic rotorcraft stability and stability in turbulence. The system defaults to 'ON' and operation is transparent to the pilot. The Force Trim System is designed to give the pilots of light- to medium-sized helicopters some 'hands free' time to perform other activities, or just to relax. The unique SASPlus (SFT-100) system combines a stability-augmentation system with a force-trim system. The HAP-100 is a digital, full-function, 2-axis (pitch/roll) autopilot system for light and medium-sized helicopters. The HFC-100 is a digital, full-function, 2-axis (pitch/roll) autopilot with a 2-axis (pitch/roll) stability and control-augmentation system for light and medium-sized helicopters, and is described here in more detail.

HFC-150 Helicopter flight control system

Included in the HFC-150, a rate-based stability control and augmentation system (SCAS) improves basic rotorcraft stability and reduces the high-rate perturbations encountered in turbulence. The SCAS series actuators, installed in the rotorcraft's primary control tubes, are very high-rate, low-authority, electromechanical actuators. The actuators operate with approximately 10 per cent control travel authority and take less than a second to move full actuator travel. For unquestioned safety, the actuators have internal electrical and mechanical stops; system architecture is shown in Figure 10.4.

Operation of the Cool City SCAS is approved throughout the full flight envelope, from start-up to shut-down, and is virtually transparent to the pilot. The SCAS System improves rotorcraft safety by reducing pilot workload and fatigue, while improving the quality of ride for all occupants.

In addition to the safety and ride-quality improvements of the SCAS, the HFC-150 provides a three-axis rate/attitude based autopilot system, increasing the pilot's ability to handle the workload extremes of the rotorcraft cockpit environment. Although the SCAS and autopilot electronics are contained within a single box, the two systems are independent. Each has their own power inputs

and internal power supplies; with a failure in one system not adversely affecting the remaining system. In addition, an autopilot failure will still leave the pilot with the option of force-trim mode to hold the cyclic in the position of engagement; therefore, the pilot still has the ability to remove his hands from the cyclic for a short period for other duties without the worry of the helicopter deviating from the intended path.

As designed, the Cool City HFC-150 will interface with essentially all models of installed attitude gyros, electronic flight instrument systems, and/or air data systems with outputs for the digital flight control system. In the event of an external failure of attitude information, the HFC-150 seamlessly switches to internal solid-state rate sensors for continued autopilot operations.

Autopilot controller

The autopilot (AP) controller – Figure 10.5 – is used to select subsystem functions and autopilot modes. It also is used to display function and mode annunciations and warnings via the LED lighting. The AP Controller is compatible for operations using night-vision equipment (see Chapter 12).

The controller measures may be installed in a single- or dual-controller configuration without

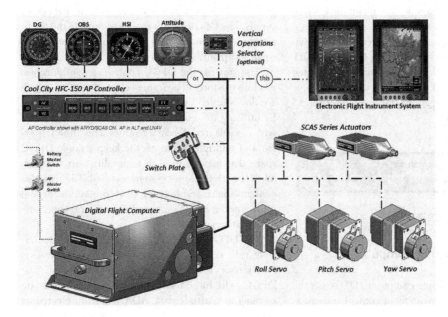

10.4 System architecture

modification to the system. The following subsystem functions are selectable from the controller:

- AP Autopilot (ON-OFF) – When AP 'ON' is selected, AP comes on in attitude-retention mode in pitch and roll axes and 'ATT' will be annunciated under HDG button
- YD Yaw Damper (ON-OFF) – Engages with AP ON – YD may be operated independently, with AP and/or with SCAS
- FT Force Trim (ON-OFF) – Only selectable when AP is 'OFF' – may be operated with SCAS and/or YD or independently
- SCAS (OFF-ON) – Defaults to ON at aircraft system power up – May be operated independently – Independent operation requires hands on cyclic at all time

The following autopilot modes are selectable from the controller:

- HDG Heading – follows heading bug from DG/HSI
- IAS Indicated Airspeed – samples and maintains current airspeed at engagement
- ALT Altitude Hold – samples and maintains current altitude at engagement. Glide slope is armed when selected while in Alt mode if tracking the localizer GS annunciation is located under ALT button
- VS Vertical Speed – samples and maintains current vertical speed at engagement
- LNAV Lateral Navigation – senses selected navigation source such as VOR, LOC, GPSS, and Back Course Back Course (B/C) annunciation is located below LNAV button
- VNAV Vertical Navigation – mode used for WAAS GPS LPV approach

10.5 Autopilot (AP) controller

Digital flight guidance computer

The digital flight guidance computer (DFGC) is the focal point of the automatic flight control system, as

10.6 Digital flight guidance computer (DFGC)

seen in Figure 10.6. The DFGC contains a variety of modern technology: surface-mount printed circuit boards (PCBs), solid-state sensors, accelerometers and air transducers. Cool City uses the latest in surface-mount technology and electronic hardware to provide high reliability and radio-frequency interference (RFI) resistance.

The DFCG makes use of a single chassis for the electronics of the system. Each box is configured for the system to be installed; however, should you choose to update the system later, this is accomplished by the addition of the requisite internal PCBs.

Although all system PCBs are located within this one box, the autopilot and SCAS are independent and each has their own power source and internal power supply. The DFGC is tray-mounted outside the cockpit; therefore, installation in the cabin area has a very small footprint.

The tray for the DFGC contains a configuration module to store all airframe-related specifics, which are downloaded by the DFGC each time at start-up. By utilizing this method, the DFGC remains generic with regards to different airframes. This allows operators of multiple helicopters to keep a single DFGC spare that may be used in many different airframe models with the same system installed. The tray also contains the air transducers; therefore, removing and replacing the DFGC does not require entry into the pitot-static system.

The DFGC contains three-axis orthogonal, solid-state rate sensors, as well as an integral yaw-axis rate sensor and a long-term slip-skid sensor. The DFGC will interface with essentially all attitude systems for attitude data, ADAHRS from electronic flight instrument systems, AHRS, and/or the air data computer that have autopilot outputs.

Any failure of external data used by the autopilot causes the DFGC to revert to the internal rate package so you do not lose autopilot functions. This reversion occurs seamlessly; however, it is annunciated on the AP Controller by the annunciation of RATE.

The DFGC has seven 'receive' and one 'transmit' ARINC 429 ports, a CAN* bus port and analogue interfaces for connecting to the aircraft systems. The DFGC also contains a digital data recorder function which records up to 48 channels of information in a 28-minute loop.

Additionally, the computer contains a bidirectional CAN port for diagnostics that can be used with an external recorder for continuous recording of operations for flight operations quality assurance (FOQA).

To differentiate between autopilot/stability-augmentation system components, Cool City uses the nomenclature of actuator for the series electro-mechanical devices used in the SAS/SCAS systems. The electromechanical devices of the autopilot system are referred to as parallel servos; however, in general, the terms actuator and servo are interchangeable.

Series actuators

Cool City manufactures two models of series actuators, a linear and a rotary series actuator. As implied in the name, the linear actuators move in and out and are installed in push-pull control tubes. The rotary actuators are used in control systems incorporating a torque tube that rotates for control travel.

The series actuator – Figure 10.7 – is manufactured on computer numerical control (CNC) machinery from solid aluminium for strength and corrosion protection. A stepper motor drives the actuator. Also, key to the structural integrity of the Cool City actuator is the dovetail mechanical joint that holds the actuator housing and the electronics box together without depending upon fasteners for strength.

---

* A CAN (controller area network) bus is a data bus standard designed to allow microcontrollers and other peripheral devices to communicate with each other without a host computer.

10.7 Series actuator

The Cool City actuators are fast-action units with very limited control travel authority. In most instances, the actuators operate full travel in less than a second and have approximately 10 per cent control travel authority. The actuators have a very high duty cycle when operating in turbulence. When quiescent or at system shut-down, the actuators automatically re-centre.

The series actuator provides SAS/SCAS actions in a short but fast motion. Mounted in the main flight-control tubes, these actuators have very limited travel authority, usually less than 10 per cent of full control travel. While limited in travel distance, these actuators operate at a high rate of speed. Full travel for the actuator occurs in less than one second.

For safety purposes, the actuators have an internal electrical and mechanical stop. The mechanical stop is located just beyond the electrical stop and is not reached in normal operation. Other than the obvious improvements in the aircraft stability, operation of the actuators is transparent to the pilot.

Parallel servo

To differentiate between autopilot and stability-augmentation system components, Cool City uses the nomenclature of servo for the parallel-installed, electromechanical devices used in the autopilot systems (Figure 10.8). The series-installed, electromechanical devices of the SAS/SCAS are referred to as actuators; however, in general, the terms actuator and servo are interchangeable.

The parallel servo is considered the most critical element in the autopilot system because it is the only

10.8 Parallel servo

component attached to the flight controls. Therefore, the servo capstan incorporates an internal spring mechanism and has an internal gradient that may be overridden in the case of a servo failure or the pilot's need to 'fly through' the autopilot.

The servo capstan gradient requires increasing force as the pilot moves the cyclic from the original override point. Upon releasing the cyclic, the spring mechanism will return the cyclic to the original position.

The autopilot servo is manufactured from a solid aluminium billet for strength and corrosion protection. For long life and smoothness of operation, the capstan rides on straddle-mounted roller bearings. Additionally, all servo gearing and shafts are manufactured from stainless steel and the shafts also run on ball or roller bearings.

Cool City uses a stepper motor to drive the servo. Stepper motors are not subject to the 'starting voltage' problems so prevalent in older systems. Also, from a safety standpoint, there is no single-point failure that can cause the stepper motor to have a 'runaway'.

The parallel servo is used in roll, pitch, and may be optionally installed in the yaw axis of rotorcraft installations. The servo is connected to the aircraft's primary control system through the use of either push-pull rods or bridle cables, dependent upon the specific requirements of the installation.

In addition to the attributes listed above, additional features include:

- 'Smart' servo design with precision servo position feedback

- Servo engage-disengage mechanism tested to 500 in. lbs
- Machined mating surfaces prevent internal contamination from external sources
- 90° electrical connector for installation in confined areas

As with all autopilot systems, the servo is most critical to optimum autopilot performance and system longevity. Cool City has successfully subjected the servo to the most strenuous environmental testing; this includes life-testing of the servo to over half a million cycles without failure.

MULTIPLE-CHOICE QUESTIONS

1. Attitude rate signals can be derived from:

   (a) Integrating the output from a vertical gyro
   (b) Differentiating the attitude output from a rate gyro
   (c) Differentiating the attitude output from a vertical gyro

2. Linear actuators are characterized by being in:

   (a) Parallel with the controls, large displacement, low gain, maximum authority
   (b) Series with the controls, small displacement, high gain, limited authority
   (c) Series with the controls, large displacement, high gain, maximum authority

3. AFCS runaway problems are most likely caused by:

   (a) Vertical gyro offset
   (b) Loss of position feedback from an actuator
   (c) Rate gyro errors

4. AFCS oscillatory problems are most likely caused by:

   (a) Rate gyro errors
   (b) Loss of vertical gyro input
   (c) Loss of position feedback from an actuator

5. Loss of the altitude-hold mode is most likely caused by:

   (a) Loss of vertical gyro input
   (b) Rate gyro errors
   (c) Loss of air-data inputs

6. An AFCS 'fly-through' feature enables the pilot to:

   (a) Make a change to the rotorcraft's flight path by disengaging the AFCS
   (b) Make a change to the rotorcraft's flight path without having to disengage and then re-engage the AFCS
   (c) Fly over a lateral waypoint at a pre-set altitude

7. Gain is the ratio of:

   (a) Input to output of a control system, which normally translates to the rate of movement of an actuator
   (b) Output to input of a control system, which normally translates to the displacement of an actuator
   (c) Output to input of a control system, which normally translates to the rate of movement of an actuator

8. Cross-coupled feedback is a technique used for:

   (a) Making a change to the rotorcraft's flight path without having to disengage and then re-engage the AFCS
   (b) Tilting the main rotor to counteract the side-slip
   (c) Faster fault detection as well as failure protection

9. The inner loop of the autopilot system is used to make:

   (a) Small flying control adjustments to counter internal/external disturbances
   (b) Large flying control adjustments to follow the pilot's control inputs
   (c) Small flying control adjustments to follow commanded guidance requirements

10. Automatic Stabilization Equipment (ASE) uses an attitude reference input to provide:

   (a) Short-term attitude-hold function
   (b) Heading-hold function
   (c) Long-term attitude-hold function

11 Autoland

Automatic landing of aircraft in low-visibility weather conditions has been made possible through parallel developments in the UK, France and the USA, leading to the ability to land an aircraft automatically in very low visibility, often referred to as 'blind landings'.

The development of airborne and ground equipment, together with crew training, led to trials being carried out on the effectiveness and reliability of fully automatic landings using the instrument landing system. In 1947, the Blind Landing Experimental Unit (BLEU) was established within the UK's Royal Aircraft Establishment. The world's first fully automatic landing was achieved in 1950. This chapter describes the development of autoland systems.

HISTORY

Low visibility is a more frequent occurrence in Europe; some examples that illustrate this are given

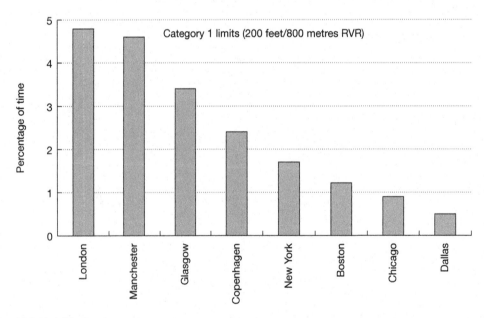

11.1 Visibility data

in Figure 11.1. The forerunners of British Airways (BEA and BOAC) took steps towards the installation of a fully automatic capability. In 1959 contracts were placed with Smiths Industries for a triplex system in the BEA Tridents, and with Elliotts for a duplex monitored system in the BOAC VC10s.

BOAC later cancelled their VC10 programme, and development work was transferred to Concorde. Meanwhile the Trident programme continued with the triplex system – Figure 11.2 – in which three independent lanes work simultaneously with a 'voting system' to identify and disconnect a failed lane. These two programmes established, through the 1960s, the processes by which the airworthiness requirements, system design, equipment specifications and supporting statistical proof of compliance should be formulated. They laid the foundations for techniques to be followed in more widespread use in many other avionic systems.

The UK airworthiness authority adopted the concept of a numerical safety level for autoland system certification, together with the need for a rigorous analysis to demonstrate its achievement. Starting from the point that the risk of a fatal accident during an automatic landing should be no greater than the risk in a clear-weather manual landing (generally accepted at that time to be 1 in 1 million), a figure of 1 in 10 million was taken as the required fatality risk per automatic landing.

Table 11.1 indicates the conversion of a qualitative severity of failure into a quantitative probability in terms of risk, to be used when autoland certification is being assessed. Having established the target risk probability, Figure 11.3 is a simplified illustration of the 'top levels' of an automatic-landing statistical model to partition the average risk between all the components. Equipment and procedures were further developed, leading to the world's first automatic landing in a passenger-carrying aircraft (the HS121 Trident) in July 1965. In 1966, the first autoland was made in a development aircraft (triplex level) in Category 3(b) conditions (0ft/50m).

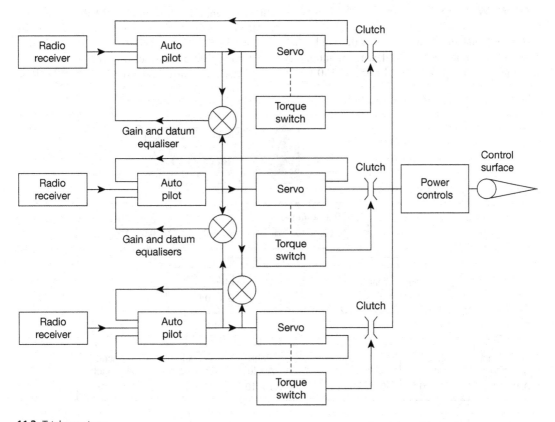

11.2 Triplex system

Table 11.1 Classification of failures

Classification	Effect	Qualitative	Quantitative
Catastrophic	Normally loss of aircraft	Extremely improbable	$<10^{-9}$
Hazardous	Large reduction in safety	Extremely remote	$<10^{-7}$
Major	Significant reduction in safety	Remote	$<10^{-5}$
Minor	Slight reduction in safety	Probable	$<10^{-3}$
None	No safety effect	None	None

OVERVIEW

In the aeroplane autopilot chapter of this book, the ILS approach mode was described, with the autopilot disengaging at a typical radar altitude of 150 feet.

Autoland can be considered as an extension of the previously described ILS coupling of an autopilot, with additional functionality:

- Radar altimeter
- Autothrottle
- Enhanced ILS beam control laws
- Automatic flare mode
- Crosswind correction prior to touchdown
- Continuation of flight guidance along the runway
- Go around
- Instruments, displays and automatic monitoring of system performance throughout the landing phase.

Design and certification of the autoland system has to consider any item of equipment failing during this critical phase of flight; failure in this context includes misleading information.

Low range radio altimeter

The low-range radio altimeter (LRRA) is a self-contained vertically directed primary radar system operating in the 4.2 to 4.4 GHz band. Airborne equipment comprises a transmitting/receiving antenna, an LRRA transmitter/receiver and a flight-deck

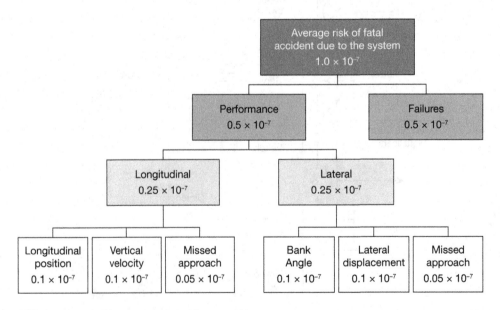

11.3 Automatic landing statistical model

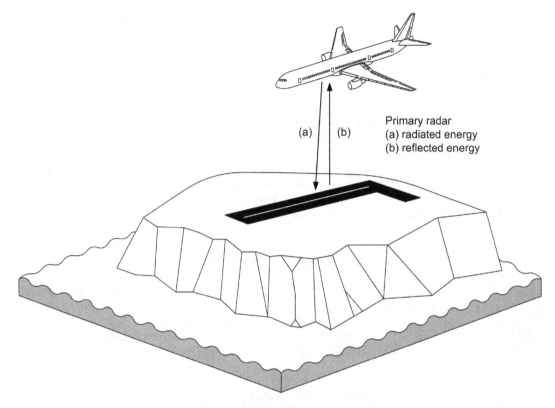

Primary radar
(a) radiated energy
(b) reflected energy

(a)   (b)

11.4 LRRA

indicator. Most aircraft are fitted with two independent systems for autoland integrity. Radar energy is directed via a transmitting antenna to the ground; some of this energy is reflected back from the ground and is collected in the receiving antenna – see Figure 11.4.

Two types of LRRA methods are used to determine the aircraft's radio altitude. The pulse-modulation method measures the elapsed time taken for the signal to be transmitted and received; this time delay is directly proportional to altitude. The frequency modulated, continuous wave (FM/CW) method uses a changeable FM signal where the rate of change is fixed. A proportion of the transmitted signal is mixed with the received signal; the resulting beat signal frequency is proportional to altitude.

Radio altitude is either displayed on a dedicated instrument or incorporated into an electronic display – see Figure 11.5. Note that radio altitude used for approach and landing is only indicated from 2,500

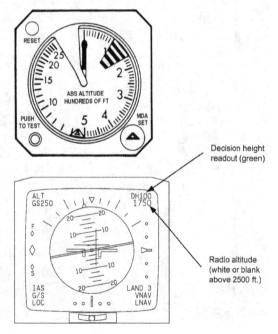

Decision height readout (green)

Radio altitude (white or blank above 2500 ft.)

11.5 Radio altitude displays

feet. The decision height (DH) is selected during ILS approaches and autoland.

Mode-select panel

Autoland is effectively an extension of the ILS approach procedures previously described. Continuation down to the runway is enabled through switch selections on a mode-select panel (MSP), as seen in Figure 11.6.

Autoland system status is normally provided by an annunciator (see Figure 11.7); there will normally be one annunciator for each pilot, connected in parallel. The autoland status annunciator (ASA) will

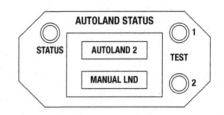

11.7 Autoland status annunciator (ASA)

only become active during the approach phase. Depending on system architecture (see later in this chapter), the annunciator will indicate the autoland configuration.

Autothrottle

Although described here as part of the autoland system, autothrottle (or thrust management system) can be used in all phases of flight, from take-off, through climb, cruise, descent, approach, flare and go-around. The autothrottle is normally used in

> **KEY POINT**
>
> The term MSP is used here; however, this item is also referred to by other names, e.g. mode-control panel (MCP).

11.6 Mode-select panel (MSP) typical location

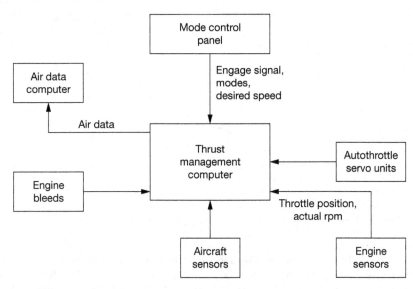

11.8 A/T system block diagram

conjunction with the autopilot's pitch channel to give a combination of speed and thrust control of the aircraft.

The autothrottle (A/T) controls engine thrust within specified engine-design parameters. The A/T is a computer-controlled electromechanical system which can either control throttle position for each engine, or (typically on older systems) each throttle lever as a group. Autothrottle controls the engine(s) to maintain either a specific engine thrust, or airspeed. Engine thrust is expressed either as a percentage of rotational speed of the low-pressure shaft, or as a function of engine-pressure ratio (EPR). The system is designed to operate primarily in conjunction with the Automatic Flight Control System (AFCS) and the Flight Management System (FMS). The desired mode of operation is selected on the mode-select panel (MSP); the engine thrust mode is selected through the FMS control display unit. Speed settings are manually selected via the MSP, depending on flight phase. When an automatic mode is selected, e.g. for an optimum climb speed as calculated by the flight management system, vertical navigation (VNAV) is selected on the MSP. The autothrottle can be used to maintain a given speed or thrust setting.

The A/T system – Figure 11.8 – consists of a computer; throttle servo mechanism(s); arm/engage switch; instinctive disengage switch; take-off/go-around (TOGA) switches and throttle lever position

transducers. Indications of A/T mode and system warnings are displayed on a flight-mode annunciator (FMA); this can either be a dedicated display unit or integrated into the electronic flight instrument system. In some A/T systems, an angle of attack (alpha) sensor located on the forward left side of the fuselage is integrated with the control laws.

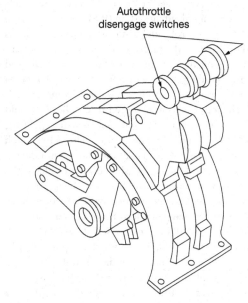

Autothrottle disengage switches

11.9 A/T disengage switches

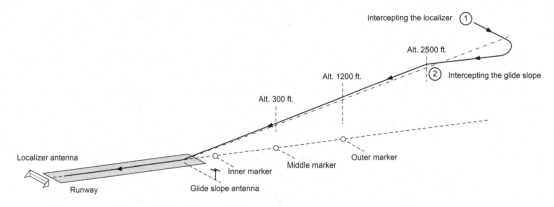

11.10 Localizer/glide slope capture

Various A/T interlocks have to be met before the A/T ARM-OFF switch will latch in the engaged position. The A/T system will disengage if a malfunction is detected, or if the throttle lever disengage switch is operated (Figure 11.9).

OPERATION

Instrument landing system

Automatic approaches are usually made by first capturing the localizer (LOC) and then capturing the glide slope (GS) – see Figure 11.10. The crew selects the instrument landing system (ILS) frequency on the navigation control panel. Runway heading also needs to be sent to the ILS receiver; the way of achieving this depends on the avionic fit of the aircraft. Desired runway heading can either be directly on a CDI, or via a remote selector located on a separate control panel.

The localizer is intercepted from a heading-hold mode on the automatic flight control system (AFCS), with LOC armed on the system. The active pitch mode at this point will be altitude hold, with the GS mode armed. Once established on the localizer, the glide slope is captured and becomes the active pitch mode. The approach continues with deviations from the centreline and glide slope being sensed by the ILS receiver; these deviations are sent to roll and pitch channels of the AFCS, with sensitivity of pitch and roll modes being modified by radio altitude. The auto throttle controls desired airspeed, with a fast/slow pointer on each attitude direction indicator (ADI); a typical electronic attitude direction indicator (EADI)

fast/slow display is shown in Figure 11.11, with F and S icons. Depending on aircraft type, two or three AFCS channels will be engaged for fully automatic landings, thus providing levels of redundancy in the event of channel disconnects.

Although the glide slope transmitter is located adjacent to the touchdown point on the runway, it departs from the straight-line guidance path below 100 feet. The approach continues with radio altitude/descent rate being the predominant control input into the pitch channel. At approximately 50 feet, the throttles are automatically retarded and the aircraft descent rate and airspeed is reduced by the 'flare' mode, i.e. a gradual nose-up attitude that is maintained until touchdown. During this flare manoeuvre,

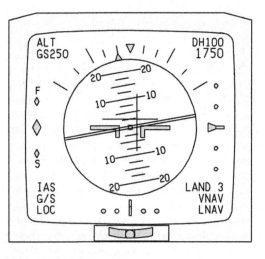

11.11 ADI fast/slow pointer

any crosswind correction is gradually reduced prior to touchdown, such that the aircraft aligns with the runway (sometimes referred to as 'kick-off drift'). The final pitch manoeuvre is to put the nose of the aircraft onto the runway. Lateral guidance is still provided by the localizer at this point until such time as the crew take control of the aircraft.

Monitoring the approach and landing

Deviation from the localizer and glide slope is monitored throughout the approach, together with confirmation of position from the marker beacons. The ILS can be used to guide the crew on the approach using instruments, when flying in good visibility. In the event that visibility is not good, then the approach is flown using the automatic flight control system (AFCS). The crew select localizer and glide slope as the respective roll and pitch modes on the AFCS mode control panel (MCP). Distance measuring equipment (DME) can also be used to determine the distance to the runway (depending on where the ground station is located). With approved ground and airborne equipment, qualified crew can continue the approach through to an automatic landing. To complete an automatic landing (autoland) the pitch and roll modes need precise measurement of altitude above

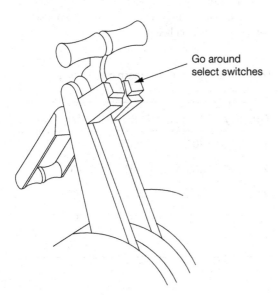

Go around select switches

11.12 Go-around (GA) switch

the ground; this is provided by the low-range radio altimeter (LRRA).

Roll-out guidance

Once the aircraft has landed, and depending on the aircraft/AFCS type, the ILS localizer provides centre-line guidance to maintain the aircraft down the runway whilst decelerating. The AFCS now controls the rudder and nose-wheel steering to maintain the aircraft on the runway centre line. On some aircraft, a paravisual display (PVD) mounted on the glare shield gives the pilot visual guidance on the landing rollout if the autopilot fails. The PVD consists of a horizontal 'barber's pole': a black spiral on a white background, this rotates in the direction of the runway centre line to guide the pilots if the aircraft drifts to the left or right of the runway centreline.

Go-around

The pilot in command can, at any time during approach and landing, elect to abandon the operation and climb away. This is normally an automatic function activated by pressing an instinctive go-around (GA) switch – Figure 11.12 – on the throttles; the pilot normally follows the movement of the throttles and will have the lower part of the hand resting on the go-around (GA) switches. There could be one or more reasons for activating GA: system/equipment malfunctions, excessive cross wind, or inability to make visual contact with the approach lights. GA mode is armed when the aircraft is established on the glide slope, and flaps are set to a landing position. A typical GA profile is shown in Figure 11.13; activating the GA switch has several distinct functions:

1. Disengages GS and LOC modes
2. Advances the throttles to a preset value of climb thrust
3. Pitches the aircraft nose up to a preset climb rate, typically 2000 feet/minute (FPM)
4. Levels the wings

Once the aircraft is stabilized in the climb, the crew will follow a pre-determined missed-approach procedure, and take further guidance from air traffic control.

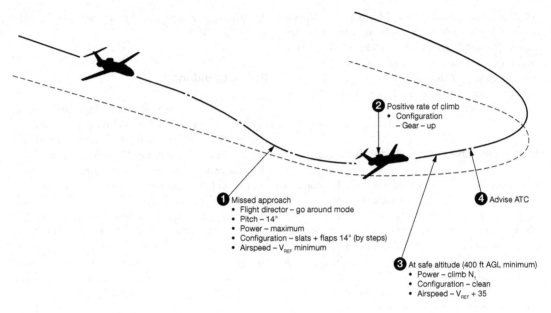

11.13 GA profile

Different aircraft types have varying GA control laws – for example, at low initial climb rates, typically less than 500 feet per minute (FPM), the pitch command will hold attitude until the rate of climb reaches 1,000 FPM when the pitch command is a combination of airspeed and attitude hold. The control laws will be adjusted in the event of an engine-out. Pitch command will always be within the aircraft's minimum/maximum speed limits, depending on flap position.

is quoted in feet, or the near equivalent in metres. Category 3 figures depend on aircraft type and airfield equipment, e.g. quality of ILS signals and runway lighting (centreline, edges, taxi ways and so on).

An operator has to have approval from the regulatory authorities before being permitted to operate their aircraft with automatic Category 2 and 3 approach and landings. This applies in particular to Category 3 decision heights.

TEST YOUR UNDERSTANDING 11.1

When is GA mode initiated?

TERMINOLOGY

Automatic approach and landings are categorized by the certifying authorities as a function of ground equipment, airborne equipment and crew training. The categories are quoted in terms of decision height (DH) and runway visual range (RVR). These categories are summarized in Table 11.2; note that RVR

Table 11.2 Automatic approach and landing categories

Category	DH (feet)	RVR (feet)	RVR (metres)
1	200	2,600	800
2	100	1,200	400
3A	<100	700	200
3B	<50	150	50
3C	None	<150	<50

TEST YOUR UNDERSTANDING 11.2

What is the difference between RVR and DH?

ARCHITECTURE

System architecture is normally designed in one of three ways:

- Simplex: Single pitch/roll/yaw channels with independent monitoring of each channel
- Duplex: Two independent pitch/roll/yaw channels with cross-monitoring between channels
- Triplex: Three independent pitch/roll/yaw channels with cross-monitoring between each of the channels

Using the above definitions, the independent channels each have their own sensors (attitude, heading, radio navigation, air data and so on), processors and servomotors. In addition, each of the channels is supplied from different aircraft electrical/hydraulic supplies. In some larger aircraft, specific flying controls are allocated to each of the channels. Depending on system architecture, any failures or abnormalities are managed in one of two ways:

- Fail-passive
- Fail-operational

A fail-passive system is able to detect a problem and automatically disconnect the failed channel without any disturbance to the flight path. Fail-passive systems are usually single channel, simplex systems with a monitoring function. There are two possible failure scenarios, either the functional channel, e.g. a pitch servomotor is driving beyond its commanded position and the pitch monitor detects the problem, or the pitch servomotor is operating normally but the monitoring function is faulty. Either way there is no automatic means of differentiating between these faults and the autopilot will disengage.

A fail-operational system is able to detect a problem and automatically disconnect the failed channel whilst maintaining automatic control of the aircraft. The system is able to continue the approach and landing unless a subsequent failure occurs. In effect, the fail-operational system now reverts to a fail-passive system.

There are derivatives and variations on these architectures, e.g. dual-dual systems where two independent channels both have their own monitoring function. A failure of one channel automatically switches in the second system and the aircraft remains under automatic control.

TEST YOUR UNDERSTANDING 11.3

Explain the terms fail-passive and fail-operational.

OPERATIONAL ASPECTS

ILS remains installed throughout the world and is the basis of automatic approach and landings for many aircraft types. Limitations of ILS are the single approach paths from the glide slope and localizer; this can be a problem for airfields located in mountainous regions. Furthermore, any vehicle or aircraft approaching or crossing the runway can cause a disturbance to the localizer beam, which could be interpreted by the airborne equipment as an unreliable signal. This often causes an AFCS channel to disconnect, with the possibility of a missed approach. The local terrain can also have an effect on ILS performance, e.g. multipath errors can be caused by reflections of the localizer; the three-degree glide slope angle may not be possible in mountainous regions or in cities with tall buildings. These limitations led to the development of the microwave landing system (MLS); see *Aircraft Communications and Navigation Systems*. MLS is not in widespread use and is not discussed further in this book.

GNSS APPROACHES

Global navigation satellite system (GNSS) is a generic term for any navigation system based on satellites; the system in widespread use today is the United States' global positioning system (GPS). Other systems in operation include the Russian global navigation satellite system GLONASS, which was established soon after GPS, and the new European system Galileo. Several nations are developing new global satellite navigation systems; at the time of writing, GPS is the only fully operational system in widespread use throughout the world. A full account of GNSS is given in *Aircraft Communications and Navigation Systems*.

Developments into the deployment of GPS have led to its use for approaches and landings, without the use of traditional radio/radar based navigation aids, e.g. VOR, ILS and DME. GPS approaches are in widespread use in the USA and are being introduced throughout Europe. There are several schemes in place or proposed to improve system accuracy, integrity and availability, including:

- Wide-area augmentation system (WAAS), maintained by the FAA.
- Local-area augmentation system (LAAS), maintained by the FAA.
- European geostationary navigation overlay service (EGNOS), a joint project of the European Space Agency (ESA), the European Commission (EC) and Eurocontrol.

Aside from being independent of any ground-based navigation aids, GNSS also offers the advantage of segmented approaches – see Figure 11.14. All these augmentation systems operate on the principle of numerous ground stations in known geographical positions receiving GPS signals. Correction signals are then sent to users in a variety of ways. The wide-area augmentation system (WAAS) was developed specifically for aviation users and is intended to enable GPS to be used in airspace that requires high integrity, availability and accuracy. WAAS improves GPS signal accuracy from 20 metres to approximately 1.5–2 metres in both the horizontal and vertical dimensions. WAAS is based on a network of reference stations around the world that monitor GPS signals and compare them against the known position of the reference stations. These reference stations collect, process and transmit this data to a master station. Updated data is then sent from the master station via an uplink transmitter to one of two geostationary satellites; the aircraft receiver compares this with GPS data and messages are provided to the crew if the GPS signal is unreliable.

A further development of GPS augmentation for aircraft is the local-area augmentation system (LAAS).

TEST YOUR UNDERSTANDING 11.4

Explain the difference between WAAS and LAAS.

This facility is located at specific airports and is intended to provide accuracy of less than one metre. Receiver stations are located in the local airport vicinity and these transmit integrity messages to the aircraft via VHF data links (VDL). The intention is for augmented GPS to gradually replace ground-based navigation aids, ultimately leading to the global navigation satellite system landing system (GLS) replacing the instrument landing system (ILS) for precision approaches and landings.

The GPS navigation receiver can also be installed with error-detection software known as receiver-autonomous integrity monitoring (RAIM). Monitoring

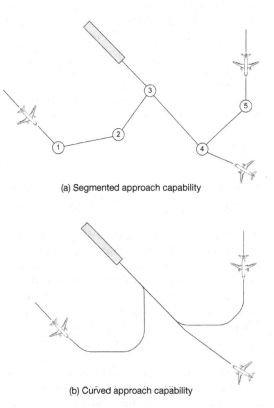

(a) Segmented approach capability

(b) Curved approach capability

11.14 GPS segmented approaches

Table 11.3 Comparison of approach technologies

Type of approach/description		
Non-Precision Approach (NPA)	Approach with Vertical Guidance (APV)	Precision Approach (PA)
Lateral guidance, but no glide slope	Lateral and vertical guidance, but not to PA criteria	Lateral and vertical guidance to ICAO criteria
Examples based on radio/radar technology		
VOR, DME, LOC	ILS	ILS
Examples based on area navigation (RNAV)		
RNAV	Localizer Performance with Vertical guidance (LPV)	GNSS Landing System (GLS)

is achieved by comparing the range estimates made from five satellites. In addition to this, failed satellite(s) can be excluded from the range estimates by comparing the data from six satellites. This technique is called fault detection and exclusion (FDE).

The introduction of GNSS-based approaches also brings new terminology; ICAO Annex 10 incorporates three classes of approach technologies – see Table 11.3

Autoland requires a combination of airborne and ground technology, crew training, operational considerations and airport infrastructure including runway lighting. These are all combined to safely guide the aircraft onto the runway at the desired touchdown point, as seen in Figure 11.15.

11.15 Final approach and landing

MULTIPLE-CHOICE QUESTIONS

1. Go-around (GA) mode is armed when the aircraft is:

   (a) On the ground roll out
   (b) Established on the glide slope, and flaps are set to a landing position
   (c) Intercepting the glide slope, and flaps are set to a landing position

2. RVR informs the pilot of:

   (a) Horizontal visibility on the approach
   (b) Vertical cloud base
   (c) Distance to go along the runway

3. RAIM is achieved by comparing the range estimates made from:

   (a) European geostationary navigation overlay service
   (b) Five satellites
   (c) Six satellites

4. Approach with Vertical Guidance (APV) gives:

   (a) Lateral guidance, but no glide slope
   (b) Lateral and vertical guidance, but to PA criteria
   (c) Lateral and vertical guidance, but not to PA criteria

5. A fail-passive system is able to detect a problem and automatically disconnect the failed channel:

   (a) Leaving the aircraft in an unstable condition
   (b) Without any disturbance to the flight path
   (c) Whilst maintaining automatic control of the aircraft

6. A category 2 approach is limited to a DH and RVR of:

   (a) 100 feet, 400 metres
   (b) 200 feet, 800 metres
   (c) 50 feet, 200 metres

7. Activating the GA switch will:

   (a) Disengage ILS modes, advance the throttles, level the aircraft in pitch and roll
   (b) Disengage ILS modes, advance the throttles, pitch the aircraft nose up and level the wings
   (c) Maintain guidance using ILS modes, advance the throttles, pitch the aircraft nose up and level the wings

8. The autothrottle is normally used in conjunction with the autopilot's:

   (a) Roll channel, to level the wings during go-around
   (b) Rudder and nose-wheel steering, to maintain the aircraft on the runway centre line
   (c) Pitch channel, to give a combination of speed and thrust control

9. A para-visual display (PVD) gives the pilot:

   (a) Fast/slow indications on the ADI
   (b) Visual guidance during the landing rollout
   (c) Vertically directed primary radar information

10. Automatic ILS approaches are usually made by first capturing the:

   (a) Glide slope and then capturing the localizer
   (b) Distance-measuring equipment
   (c) Localizer and then capturing the glide slope

12 Electronic display technologies

Electronic display technologies in widespread use are synthetic vision and enhanced vision; both are designed to enhance the pilot's situational awareness.

Synthetic vision is a computer-generated system that uses three-dimensional (3D) images to provide pilots with clear and intuitive means of understanding their environment. Using sophisticated graphics modelling, synthetic vision tracks the navigation system's terrain-alerting database to recreate this 'virtual reality' landscape on the primary flying displays (PFDs). Synthetic vision can include images of: ground and water features, airports, runways, obstacles, traffic and so on.

An alternative technology is enhanced vision systems (EVS), typically based on infrared (IR) sensors, to provide vision in low-visibility conditions. The output from an IR sensor mounted in the nose of the aircraft is processed and then projected as an image, typically on a heads-up display (HUD). Objects detected by the IR sensor are proportional in size and in perspective with features outside the aircraft. In low visibility the pilot is able to view the IR sensor image, then to transition from on-board guidance displays to the external environment as the aircraft approaches the runway.

Night-vision imaging systems (NVIS) are based on enhanced vision, using night-vision goggles (NVG); this is a powerful technology for aircrew during low-light/night-time operations. Based on military technology, NVIS is also employed in civil aircraft, typically police and medical rotorcraft. NVIS allows the pilot/observer to look outside the aircraft at night, enabling them to perform safer take-offs and landings and (for specific mission purposes) operate closer to the ground.

SITUATIONAL AWARENESS

Defining 'situational awareness' is not an easy task because the phrase means many things to many people. The author's definition is '*the aircrew's continuous perception of the aircraft in relation to the dynamic environment of the flight profile, including other traffic, terrain and so on, and the ability to forecast and subsequently execute tasks based on that perception*'.

Loss of Situational Awareness is a major contributor to aircraft accidents; CMC Electronics Inc. has published the following statistics for business jet accidents:

- 62.9% of accidents attributed to loss of situational awareness
- 41.4% of accidents occurred during approach and landing phase of flight
- 48.1% of 27 controlled flight into terrain (CFIT) accidents cited loss of situational awareness as contributing factor
- 64.4% of approach and landing accidents (ALA) involved lack of stabilized approach
- 10% of accidents involved hard landings

Synthetic vision and enhanced vision are both designed to enhance the pilot's situational awareness. There are advantages/disadvantages to both synthetic and enhanced vision; this book is not making a case for either technology. Both systems are described in more detail in this chapter.

SYNTHETIC VISION TECHNOLOGY

One of the industry leaders of synthetic vision is Garmin with synthetic vision technology (SVT™); this provides the pilot with a realistic 3D view of topographic features surrounding the aircraft – see Figure 12.1. Using sophisticated graphics modelling, SVT recreates a visual topographic landscape from the system's terrain-alerting database. The resulting virtual-reality display offers pilots a supplemental 3D depiction of ground and water features, airports, obstacles and traffic – all shown in relative proximity to the aircraft.

An early chapter described the Garmin 1000 electronic primary flying display (PFD). In this system, SVT is integrated with the PFD; SVT is a virtual revolution in 'big-picture' visualization. Instead of a flat 'blue-over-brown' attitude/heading reference, the pilot sees an in-depth perspective view of realistic terrain features rising into the sky. The pilot 'looks through' the PFD to monitor the flight situation – without having to piece together a mental picture solely from raw instrument data on the panel.

When flying in areas or at altitudes where rising terrain may pose a hazard, SVT uses its terrain-alerting database to colourize the landscape – clearly showing with amber or red overlays those areas where potential flight-into-terrain risks exist. Any towers or obstacles that may encroach upon the flight path are

(a)

(b)

(c)

(d)

12.1 Synthetic vision technology (SVT™): (a) Pathways (b) Obstacles (c) Traffic (d) Runway

colour-highlighted and clearly displayed with height-appropriate symbology.

SVT shows fixed terrain and obstacles, and also works with various traffic advisory / warning technologies to depict targets in 3D perspective, so the pilot can visually gauge how high and how close the other traffic is. In addition, the familiar colour – and shape-cued traffic symbology grows larger as it gets nearer – makes traffic conflicts easier to see and identify.

In addition to identifying airports and showing runways in graphical perspective, SVT also helps simplify en-route navigation. It can create a three-dimensional 'pathway' view on the G1000 PFD, showing en-route legs, terminal procedures, and ILS or GPS/WAAS vertical approaches – all laid out in front of the aircraft by means of outlined 'windows' on the display. These windows vary in size to depict the flight path in perspective; guidelines in each corner of the pathway windows point in the direction of the active flight plan leg – making it easy for pilots to follow the 'pathway in the sky'.

Although Garmin SVT is not intended to replace traditional attitude and directional cues as the primary flight reference, it clearly augments the pilot's view of this data by giving it a realistic visual frame of reference – so the pilot can see at a glance where the aircraft is in relation to ground features, traffic and the flight path entered in the navigation system.

ENHANCED-VISION SYSTEMS

Enhanced-vision systems (EVS) provide vision in limited-visibility environments; these images can be integrated with flight displays, as shown in Avidyne's Entegra system – see Figure 12.2. EVS typically employ nose-mounted infrared (IR) sensors,

12.2 Avidyne Entegra EVS

operating in the electromagnetic spectrum (Figure 12.3). EVS images can also be integrated with a head-up-display (HUD), and on multifunction displays. The IR images on the HUD are the same size and aligned with features outside of the aircraft. In low-visibility conditions during an automatic approach and landing, the pilot is able to view the IR camera image and transition to the external environment as the aircraft approaches the runway. During the approach, the pilot can observe the familiar runway environment – e.g. approach lights, runway markings and so on – in preparation for the landing. Forward-looking infrared (FLIR) is a technology detecting objects by their long-wave infrared temperature emissions.

CMC Electronics Inc. has developed the SureSight® enhanced-vision system (EVS) family of products to increase flight crew situational awareness by helping them see through fog, haze, precipitation and at night to improve overall safety and aircraft economic efficiency. This is accomplished by using infrared (IR) and millimetre wave radar (MMWR) sensor systems.

EVS provides an image on the head-up display (HUD), a head-down display (HDD) or both to enable the pilot(s) to see the terrain/airport environment in low-visibility situations. EVS significantly improves situational awareness, not only during take-off, approach and landing, but also during ground manoeuvring.

HEAD-UP-DISPLAY

Head-up-display (HUD) technology was developed for military applications, and is now in use in civil-aviation applications, particularly business aircraft. The advantage of a head-up-display system stems from the collimated image, i.e. an image focused at infinity. Symbols depicting various cues are generated on an electronic display/projector located in the cockpit roof (see Figure 12.4). These symbols are projected onto a glass plate in the pilot's sightline, the combiner.

Collimation is an optical technology that produces parallel light rays. The pilot's eyes can focus on infinity to get a clear image at or near optical infinity. Typical

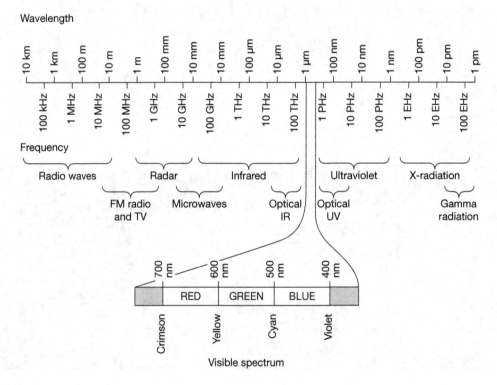

12.3 Electromagnetic spectrum

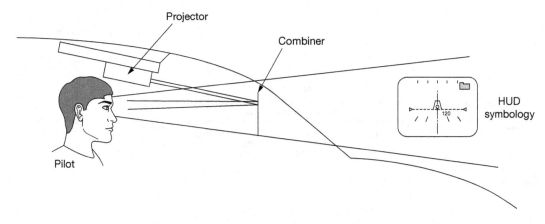

12.4 Typical HUD system: cockpit arrangement

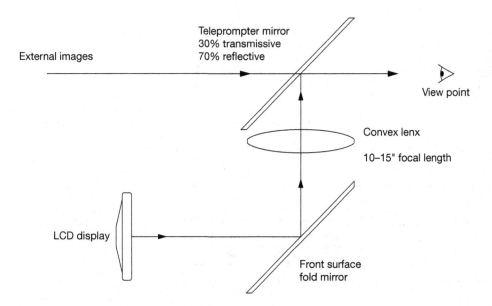

12.5 LCD and mirror arrangement

HUD technology based on an LCD display is illustrated in Figure 12.5. The combiner is transparent material that displays data and symbology allowing the pilot to view information with the eyes looking forward and out of the cockpit, instead of 'head-down', looking at instrument-panel displays. Computer-generated symbology is superimposed on the external environment in a combining effect. The projected image remains aligned with the external environment, irrespective of the pilot's head/eye movement.

The benefits of a HUD system include:

• Display of visual information overlaid on the external environment
• Essential flight data displayed in a localized field of view
• Improved situational awareness

A typical HUD system – Figure 12.6 – comprises a guidance computer, optical projector and combiner. The optical projector is installed in the cockpit roof;

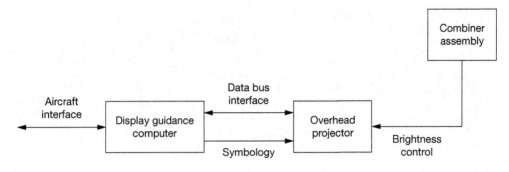

12.6 Block diagram

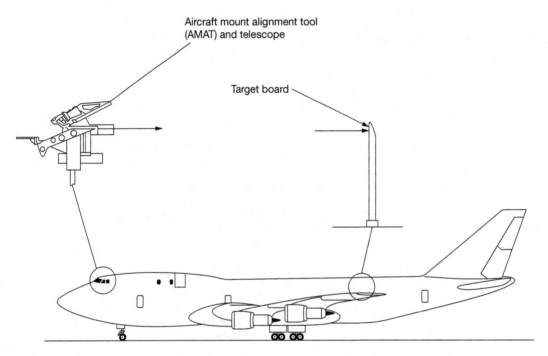

12.7 HUD bore-sighting

the combiner is attached to the overhead panel, in the pilot's field of view. The combiner is folded away when not in use.

Bore-sighting is required on HUD components in the cockpit to align with the aircraft's three axes. This ensures that the HUD images are referenced to the earth's horizon and the aircraft's projected flight path (Figure 12.7). When calibrated, the display of visual cues, e.g. the runway, is aligned when the runway become visible.

Optical projection in early, first-generation, HUD systems was based on a cathode ray tube (CRT).

Second-generation systems use a solid-state light source, for example a light-emitting diode (LED) or liquid-crystal display (LCD), to display the images. Next-generation systems use a variety of technologies including optical waveguides and scanning lasers to produce images directly in the combiner rather than using a projection system.

HUD symbology replicates the critical flight parameters needed for an approach and landing, e.g. airspeed, altitude, attitude, heading and so on – see Figure 12.8. Other information (e.g. attitude, heading and so on) is displayed in analogue format. HUD symbology matches the PFD in terms of colour and shape to aid transition from the head-down to head-up perspectives. An important feature of the HUD is the flight-path vector superimposed on the outside environment; this provides the pilot with an instantaneous and intuitive indication of the direction and speed of the flight profile.

Symbols and data used in typical HUD systems include:

- Bore-sight
- Flight-path vector
- Velocity vector
- Acceleration indicator
- Energy cue
- Angle of attack indicator
- Navigation data and symbols

The bore-sight symbol indicates where the nose of the aircraft is directed. The flight-path-vector (FPV) or velocity-vector symbol indicates a predication of the aircraft's flight path, e.g. if the aircraft is pitched nose down, but is losing energy, the FPV symbol will be below the horizon even though the bore-sight symbol is above the horizon. During approach and landing, the pilot maintains the FPV symbol at the desired glide-path angle and touchdown point of the runway. The acceleration (or energy cue) display indicates if the aircraft is accelerating or decelerating. The angle-of-attack indicator displays the wing's angle relative to the airflow. Various navigation data and symbols are used during approaches and landings, integrating ILS and/or GNSS sensors.

Using the flight-path vector and other symbology allows the pilot to control the aircraft manually and with greater precision compared to head-down

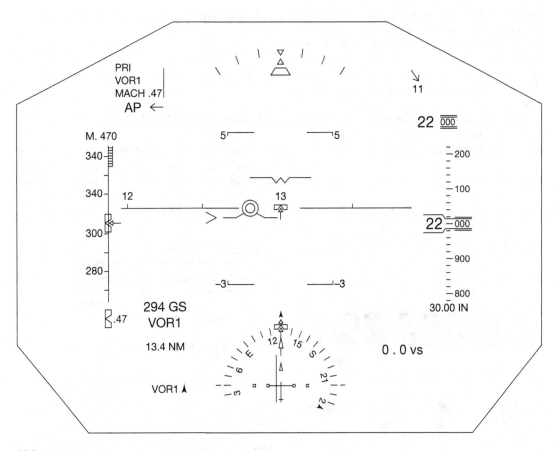

12.8 HUD symbology

displays. In low-visibility approaches, pilots can fly head-up using vertical and lateral directional cues, either from the ILS or from the GNSS. The pilot's focus is literally the point needed to visually acquire the runway lights at the approach minima. The HUD will be integrated into the flight guidance system to perform a missed approach if necessary.

The flight-path vector can also be integrated with the terrain-awareness warning system (TAWS) and traffic-collision avoidance system (TCAS), displaying the escape manoeuvre as required. Both the escape manoeuvre and search for terrain or traffic can be accomplished using the head-up display, e.g. during reduced-visibility take-offs.

Integration of the HUD and the overall cockpit environment has to be investigated and developed for new and retrofit applications. An integrated transport aircraft cockpit from CMC Electronics is shown in Figure 12.9. This level of integration is driven by human factors engineering (HFE); this ensures that operators are able to accommodate, adapt to and interface with a complex suite of new equipment, such that role performance and mission effectiveness may be enhanced. The net effect is one of maximizing overall system effectiveness within the constraints imposed by human, technological and/or cost considerations.

HUD applications for rotorcraft are illustrated by CMC's HeliHawk (Figure 12.10). HeliHawk features include a single adjustable combiner, auto brightness

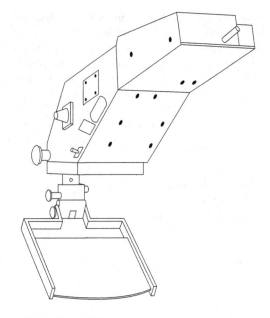

12.10 Helihawk HUD

control, electronic bore sighting, continuous built-in test (BIT) and night-vision goggle (NVG) compatibility.

COMBINED VISION SYSTEMS

Combined vision systems (CVS) integrate two or more of the following technologies in various ways, such that the capability of the whole system may exceed the sum of the parts. The various technologies are shown in Table 12.1:

Table 12.1 CVS technologies

Term	Technology	Based on
EVS	Enhanced-vision system	External real-time imaging
SVS	Synthetic-vision system	Computer-generated imaging
HUD	Head-up display	'Out of the window' optical images
HDD	Head-down display	Cockpit displays
EFB	Electronic flight bag	Cockpit displays

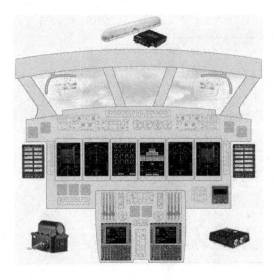

12.9 HUD/cockpit integration

The CVS concept involves a variable combination of synthetic and enhanced systems. A typical example would be database-driven synthetic vision images combined with real-time sensor images superimposed and correlated on the same display. This includes selective blending of the two images based on the intended function of the combined vision system. For example, on an approach, most of the arrival would utilize the SVS images; as the aircraft nears the runway, the picture transitions from synthetic to enhanced vision, displaying the runway and its environment to provide situation awareness.

The ultimate aim for CVS is to provide reliable and consistent all-weather operations, including zero-visibility landings and ground manoeuvring. Although Category III approach and landings are well established, using CVS the limitations are what the aircraft is equipped with, not the ground infrastructure.

NIGHT-VISION IMAGING SYSTEMS

Introduction

Night-vision imaging systems (NVIS), used in conjunction with night-vision goggles (NVG), are a powerful aid to aircrew during low lighting, night-time operations – see Figure 12.11. Based on military technology, NVIS is also employed in civil aircraft. Light can be characterized either as particles (photons) or as electromagnetic energy; both principles are used when describing NVIS. NVGs are essentially image intensifiers that amplify low light levels such that they can be displayed to the flight crew in monochromatic images. The NVG utilizes a specific range of the infrared spectrum that is not entirely visible to the human eye. In basic terms, the NVG amplifies red and IR light and then electronically transposes the

12.11 Airborne NVG

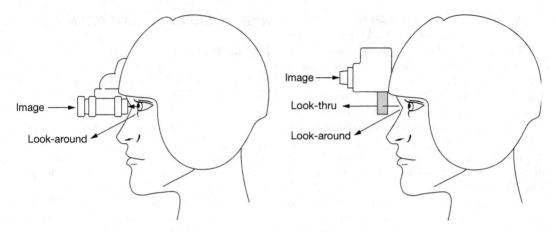

12.12 NVG: (a) Type I; direct vision (b) Type II; indirect vision

resultant images to the visible spectrum by displaying images on a green phosphor screen.

NVIS can either be viewed directly (Type I) or indirectly (Type II) – see Figure 12.12. The evolution of NVG has been developed through various generations, from ground use by the military through to airborne applications. These NVG generations have evolved as size and weight have been decreased, combined with increases in performance.

NVIS allows the pilot or observer to look outside the aircraft at night, enabling them to perform safer take-offs and landings and (for specific mission purposes) to operate closer to the ground. Typical civilian missions using NVIS include search and rescue, medical evacuation, forest-fire patrols and police surveillance operations. NVGs operate in the near infrared (IR) part of the electromagnetic spectrum; this IR energy is reflected from stars, the moon and man-made, cultural lighting. One major consideration for airborne use of NVIS is that it has to account for the lighting inside the cockpit; incandescent lighting emits more than 90 per cent of lighting energy in the IR frequencies. Since the NVGs are tuned into IR, incandescent lighting would saturate the NVG and render them useless to the pilot/observer. The solution is to fit NVG-compatible lighting in the cockpit and (where applicable) in the cabin. This can be achieved with filters over incandescent lighting, or by installing lighting that only emits in the IR range. NVG technology is based on monocular or binocular sensors. Monocular sensors have no depth perception, dual-sensor systems provide normal vision and full depth perception to the wearer.

Colour perception

Colour, as perceived by humans, results from a given spectral distribution of light, detected by three colour sensors in the eye and then interpreted by the brain. With emerging electronic display technologies in the 1920 and 1930s, experimentation and mathematics enabled the response of the human eye to be characterized. An international group, the International Commission on Illumination (CIE), established three CIE colour-matching functions, that (for practical applications) can be treated as if they were the approximate spectral response curves for the human eye in red, green and blue – see Figure 12.13. The human eye has a tristimulus characteristic in these three regions of wavelength: short (blue), medium (green) and long (red).

Scotopic vision occurs under low-light/night-vision conditions. The human eye is most sensitive to wavelengths around 498 nm (green-blue) and is insensitive to wavelengths longer than about 640 nm (red). Photopic vision occurs in well-lit daylight conditions; the human eye allows colour perception, and a significantly higher visual acuity and temporal resolution than scotopic vision. The relative responses of the eye under scotopic/photopic vision are given in Figure 12.14.

In 1931, the CIE published the chromaticity diagram (see Figure 12.15), which plots the entire gamut of human-perceivable colours by their 'xy' coordinates. Referring to the chromaticity diagram, it can be seen that the monochromic wavelengths are wrapped around the edge of the gamut. (Author's

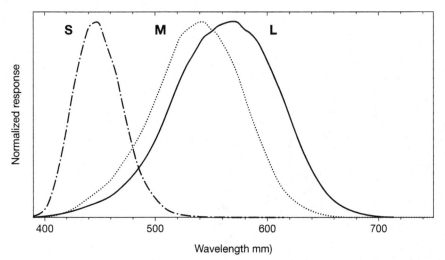

12.13 Colour-response curves for the human eye

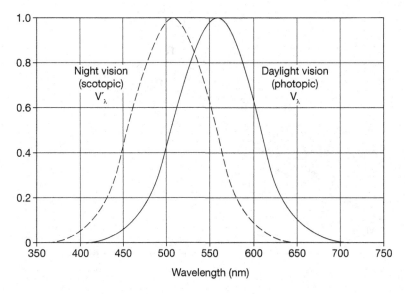

12.14 Scotopic/photopic vision

note: the mathematical development of the chromaticity diagram is beyond the scope of this book. The relationship between wavelengths and intensity can only be fully appreciated in colour; there are many images of the CIE chromaticity diagram available on the internet.)

It is a requirement for ANVIS use for the cockpit lighting of every night system to be modified to adapt to the special colour and low-intensity requirements of this system.

The allowable night-colour spectrum is essentially restricted to the range between violet/blue and yellow, with most lighting implemented in green,

centred on 550nm. Standard white or blue/white cockpit illumination is not permitted during ANVIS operations; so the changes can be significant to an existing cockpit. This colour restriction creates some problems when amber and red indicators must be present for safety reasons.

Even at low levels, colours from amber to red can cause stray emissions of these colours that can effectively ruin night-vision operations. Halo effects, known as 'blooming' in goggles, can override the image that is being viewed. Blooming refers to distorted NVG images surrounded by obscuring halos of light caused by optical saturation of the goggles. Yellow

12.15 1931 CIE chromaticity diagram

is the ANVIS warning colour, but of course it still looks green if seen through goggles. Colours, such as red, are permitted in some installations based on the low likelihood of activation during normal operations; suppression of their infrared signatures must be done, or the resulting bloom in the goggles will make them unusable.

Cockpit colours have to be changed to the blue/green region for NVIS compatibility if the flight crew looks down at the instruments without NVG. This colour change, or shift, must be done because the goggles provide very great light amplification, typically 10,000 to 50,000 times or more. As a result, even small amounts of trace or reflected light in the red to infrared passband of the goggles seriously degrades goggle operation; it causes blooming and severe visual distraction for aircrews. Even a single glance down with incompatible cockpit lighting would 'blind' the wearer, compromising safety.

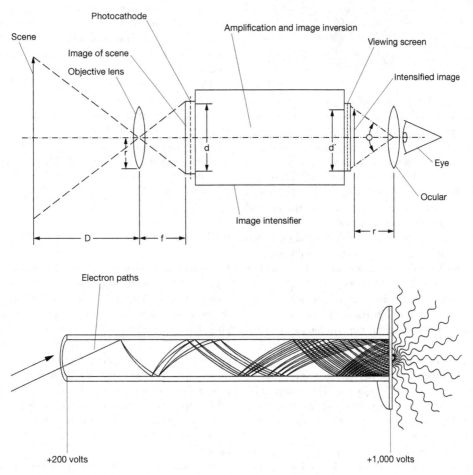

12.16 Image intensifier principles: (a) Features (b) Converting photons into electrons

Image intensifier

The basic building block of the NVG is an image-intensifying tube (IIT, or I²T) with high sensitivity to nearby infrared light as well as some visible light – see Figure 12.16.

> **KEY POINT**
>
> Infrared light has longer wavelength compared to visible red, above 750nm; human vision usually stops at about 680nm.

> **KEY POINT**
>
> Passband is the optical wavelength window (from x nm to y nm) that permits light to pass.

Near infrared light is made of longer, normally invisible wavelengths beyond 750 nanometres. All of these wavelengths are collected and amplified, then translated to a single common shorter visible wavelength that human eyes can see. In this way, an image (usually green in colour) is formed from the wideband light energy that is present, but either too low in level to be useful to the human eye or outside the normal wavelength range of human vision.

While they can allow significant night-vision improvements (up to 50,000 times light amplification), IIT have a limited tolerance for unwanted or stray light, or light in the wrong colour spectrum. These issues create integration problems when goggles are used in an 'off-the shelf' cockpit design, and using goggles effectively requires careful planning.

This situation is further complicated by the fact that goggle use is typically only 10 to 20 per cent of total flight time for most aircraft; therefore, cockpit designs must be both daylight and NVG compatible, regular night-flight compatible and satisfactory to EASA and the national aviation authority (NAA) from a safety standpoint.

NVG are designed to operate in very low ambient light but will not work in total darkness. Each IIT has a built in automatic control that reduces the efficiency of the intensifier as the ambient light increases. This reduction in efficiency results in decreased resolution and therefore decreased visual acuity. NVG are used to view the outside world and are collimated at infinity. As there is no automatic refocusing device, the inside of the cockpit is viewed with the naked eye and must be illuminated with NVG-compatible lighting.

The image intensifier has high sensitivity to near infrared (IR) light as well as some visible light. Near IR light is made of longer, normally invisible wavelengths beyond 750 nanometres.

All of these longer wavelengths are collected and amplified, then translated to a single common shorter wavelength our eyes actually can see. In this way, an image (usually green in colour) can be formed from the wideband light energy that is present but either too low in level to be useful to the human eye or outside the normal wavelength range of human vision.

Very high optical gains in these later generation systems can give rise to operational problems in the cockpit, since any stray light saturates, or overloads the display. The wide visual dynamic range is difficult to deal with.

The spectral content above 700–750nm, which is present in starlight, is not visible to the unaided human eye. NVIS systems convert this available energy to shorter wavelengths the eye can see – allowing vision under circumstances that would normally appear 'dark' to the unaided human eye. Relative responses for NVIS and the human eye are given in Figure 12.17.

NVIS evolution

First-generation (Gen I) NVIS predated their use in aviation; being used by ground-based military or security forces; these are based on an optical gain in the order of one to two thousand, projected onto a phosphor green display. Second-generation (Gen II) NVIS based on a micro-channel plate gave an enhanced image intensifier, together with higher quality optical components, leading to an even higher gain and clearer images. Third-generation (Gen III) NVIS introduced improved photo-cathode materials. Military specified Gen IV systems have even higher optical gain (in the tens of thousands) and enhanced operation.

Airborne or aviation NVIS (ANVIS) generally allow the user to look down without the goggles to

Relative responses of night vision goggles and the human eye

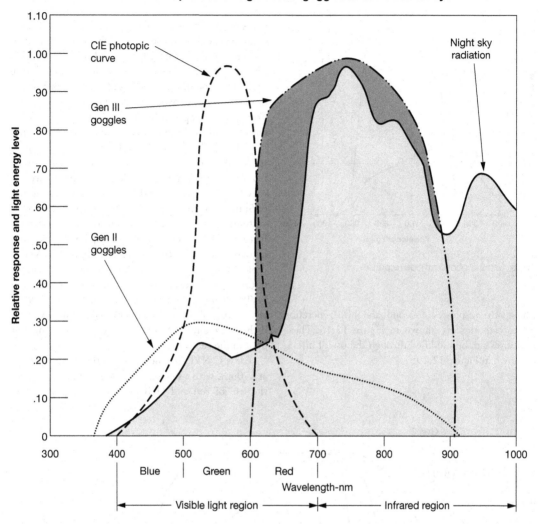

12.17 Relative IR responses (eyes vs NVG)

examine the instrument panel, then forward through the goggles to see outside terrain, thus avoiding some light overload and focus problems.

Aviation-use NVGs are binocular devices with each body consisting of an optical system and an image intensifier. The intensifier may be powered from a single self-contained power pack or each body powered individually. The optical system is designed to focus the image onto the image intensifier and to enable the operator to view the intensified image. They operate from 625nm (or 665nm) to 900nm spectrum.

KEY POINT

Terminal voltage of internal batteries on self-powered systems will reduce in flight, potentially causing unexpected loss of night vision at a critical time.

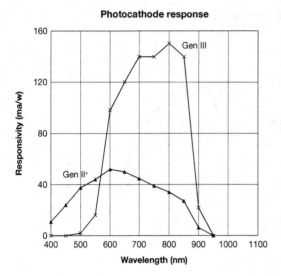

12.18 Gen 2/3 photocathode response

The relative responses of second- and third-generation photo cathodes are shown in Figure 12.18. These responses can be modified through the use of filters, as shown in Figure 12.19.

Classification of NVIS systems

Class A NVIS has spectral content starting at 625nm, enabling maximum use of ambient light to identify terrain features, ideal for low-altitude operations. This 625nm cut-off makes the use of red and full-colour multi-function displays impossible (as anything in the orange-red region would saturate the goggles), but it provides high-optical sensitivity in flight. The compatible cockpit illumination, or NVIS radiance for class A, is called NRa. Use of Class A NVGs can lead to certification issues due to the requirements to maintain red warning indications, e.g. engine fire.

Class B NVIS is when the spectral content starts at 665nm, allowing wider colour use in the avionics systems, but with reduced sensitivity of external features. This is achieved by an objective lens filter incorporated into the NVG. This class of NVG is typically used where low-altitude terrain following is not required. The compatible cockpit illumination, or NVIS radiance, for class B is called NRb. Class B NVG are compatible with red cockpit lights, and are used if there are any NVIS lighting components emitting light in the red range.

Class C NVIS is when the spectral content starts at 670nm, with a secondary peak at 540nm, allowing the use of red and HUD systems working in this

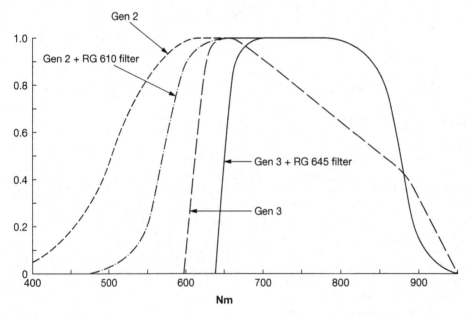

12.19 Gen 2/3 with filters

colour range. The compatible cockpit illumination, or NVIS radiance, is called NRc.

NRa systems generally operate with all types of NVG; however, NRb systems can interfere with Class A goggles. Class C goggles are very specialized and require a custom light environment. The relative response of NVIS A, B and C is shown in Figure 12.20.

KEY POINT

Class B aircraft can have NVIS lighting components which emit light in the Red range, as required to meet the specifications defined in RTCA/DO-275 and EASA/ FAR requirements.

KEY POINT

Class A aircraft cannot have any NVIS lighting components which emit light in the red range.

Day/night modes

Day warnings and cautions are normally presented at full brightness. Instrument and panel lighting are extinguished. Night warnings and cautions are normally presented at brightness clearly discernible for night operation, depending on pilot preferences. If a dimming capability is provided, all annunciators,

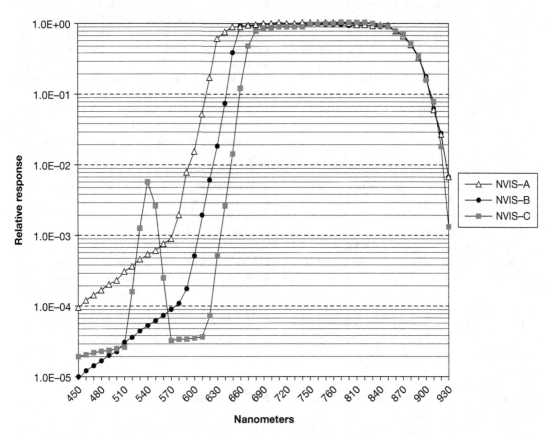

12.20 Class A, B and C curves

including master warning and caution, may be dimmable as long as the annunciation is clearly discernible for night operation at the lower lighting level. Undimmed annunciations have been found unacceptable for night operation due to disruption of cockpit vision at the high intensity.

Warnings and cautions are presented at a fixed luminance of 15 foot-Lamberts (fL), which maintains attention-getting capabilities whilst not degrading the operation of the NVIS. Instrument and panel lighting is variable from extinction to full brightness and any non-NVIS-filtered equipment lighting is extinguished.

Prior to installing to an aircraft, any NVIS-compatible equipment should be viewed in a darkroom facility, both with a test set and the selected NVGs. During this assessment any unfiltered light will be detected and the effect of the lighting on the performance of the NVG will be established. The testing should be carried out by suitably qualified engineers who have experience of NVIS-compatible lighting and who are able to recognize the full range of effects due to unfiltered light sources.

Following build of the NVIS aircraft, or after an upgrade, the aircraft windows are blacked out to simulate a dark night ambient lighting condition. This may require that the hangar itself is darkened. The whole cockpit lighting installation, and where applicable cabin/equipment lighting, can then be assessed for:

* Readability of instruments, controls and displays
* Lighting balance of self-illuminated equipment panels and displays
* Attention-getting/brightness/balance of warning and caution indicators
* The effect of reflections in the transparencies on the view out of the cockpit
* Compatibility with the selected NVG
* The effect of reflections on the selected NVG
* Cockpit ergonomics

To use ANVIS systems, the cockpit must be modified to provide compatible low-light levels at the correct optical wavelengths; if not, the residual light will overwhelm the goggles, making them unusable. This cockpit emission is called '(A)NVIS radiance' (AR) and is the total optical emission from 450nm to 930nm. The specific colours within this range are defined as a given chromaticity.

The term 'NVG-compatible' has several definitions; in this book it is defined as 'where internal and external lighting does not degrade the NVG image'. NVIS is a system that combines a selected NVG with compatible interior and exterior illumination sources.

Ensuring that a cockpit environment is NVG compatible is a complicated integration task, as many different systems (from different manufacturers and with different dynamic performance) need to be brought into some kind of optical harmony for dimming to low levels and correct coloration/chromaticity and IR suppression.

While there are approved NVIS colours for green, yellow and red (all with highly reduced IR signature), it is important to realize that they all look the same when goggles are used – all appearing in various shades of green to the eyes of the flight crew. For this reason, most NVG allow the aircrews to look down, below the actual image intensifier, to see the instruments (and their original colours) directly.

From the response curves, NVGs are about 1,000 to 10,000 times less sensitive to light in the 450 (blue) to 600 (orange-red) range than in their design target of about 750nm in the IR range. As light moves into longer wavelengths (red region), the difference drops to a factor of 10 to 100, which explains why goggles are seriously affected by even a small amount of light in the wrong spectral range. To achieve compatible operation, it is critical to suppress emissions in the red to IR range from cockpit displays, lights and equipment as much as possible. This is done by selection and design or through optical filtering of the light emitters.

Unfortunately, the eye is not a very useful tool for making this judgement. For example, things that look green (including LEDs) also could contain significant IR energy at the same time, which the unaided eye cannot detect. This unwanted IR signature manifests as bright blooming (and subsequent loss of sensitivity) in ANVIS goggles. While blooming is easy to see, it is not so easy to characterize. Many people struggle with coloured filters (typically green) to achieve NVIS compliance, but often this is inadequate, leading to many cockpit problems during goggle use because of unsuppressed high IR emissions even in green-coloured displays.

NVG compatibility

When looking at lamps, filters, panels and other illuminated objects, cockpit instrument optical energy emitted must be within the NVIS radiance range and chromaticity compatible with the type of aircraft and ANVIS system. For rotorcraft, the key is with NRa or Class A NVIS systems and compatible illumination.

Generally speaking, emission must be suppressed above 600nm (red and beyond), and acceptable emission normally peaks at about 530–550nm (aqua to green). Illumination filters identified as NVIS Green A for normal lighting and NVIS Yellow A for alerts in a helicopter.

Colours and filtering

Optical filters often can be found with a white illumination source (either LEDs or incandescent lamps).

This technique originates from filtering 'white' incandescent lamps, and it is the least efficient way to achieve NVIS compatibility today.

Light can be specified by its quality of colour (termed chromaticity) in terms of hue and intensity. Relatively speaking, white LEDs have significant IR signatures and poor energy content in the desired green region. The use of deep green LEDs eliminates the unwanted IR signature and provides nearly perfect chromaticity and radiance with virtually no secondary filtering.

In 1976, the CIE chromaticity diagram was updated based on continued research and development of colour technology (see Figure 12.21). Three NVG-compatible colours are defined by this diagram: NVIS blue, green and red. These colours are defined by their coordinates on the diagram.

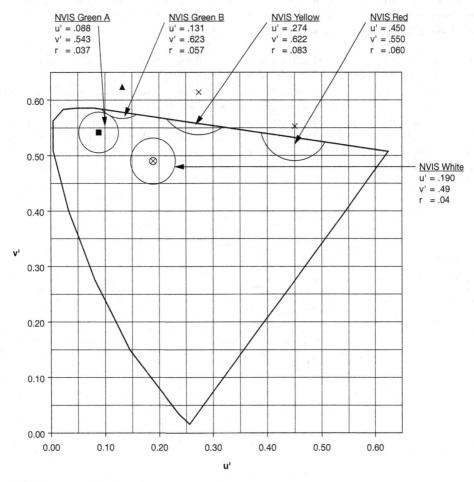

12.21 1976 CIE chromaticity diagram

Panel or instrument lighting that is intrinsically NRa-correct consumes less power, is less expensive to build, is more reliable and is easier to implement than a complex filtered system.

The use of broadband optical filters (originally designed for plasma displays) makes it possible to suppress IR emissions from any type of optical display, (e.g. CRT or LCD) while retaining normal operation and colour. These filters can be useful in converting a cockpit to an ANVIS-compatible environment.

To aid in analysis and correction of lighted cockpit systems, an IR passband filter (which passes IR but blocks visible light) can be used in front of the test item; 720nm is a good filter choice as it passes virtually all IR emissions.

Backlit LCD displays can be especially difficult to filter adequately; however, neutral grey IR filter films and materials are readily available to solve this problem. Note that coloured optical filters might not have IR-blocking ability, despite the change in visible colour.

While LEDs generally are narrowband emitters (except for white), colours of yellow through green might contain unexpectedly large amounts of IR emission; even green LEDs can have enough IR emission to be unacceptable due to the high image-intensification factor of the NVG. This principle is termed 'minus IR' filtering, and is illustrated in Figures 12.22 and 12.23.

Dimming

One common problem in an NVIS-integrated cockpit is that systems can dim in various ways. This problem can often be improved by adding series resistors (adequately rated for power) from the common dim bus to individual items that are simply too bright. Series resistance, diodes or even power zener diodes (to create a large dimming voltage gap) can all serve to equalize the distributed bus to specific problem units.

Matching NRa levels is a complicated task and requires a broadband NVIS radiometer (optical power meter) with sensitivity corrected for the blue-green spectrum. Most of the optical test instruments widely available are for fibre-optic-based measurements, which are all red to infrared, and thus have low, or no, blue-green characteristics. Some optical-spectrum analysers and NVIS radiometers work over the full required range and provide good analytical tools for the design and characterization of NVIS systems.

Faults with NVIS-compliant hardware can manifest as blooming, glare and significantly reduced sensitivity of the goggles because of light interference. Impaired external vision can be a serious safety issue, and the cockpit cannot be accepted as compliant until the flight crew feels there is no degradation in this area, regardless of measurements, standards or test data.

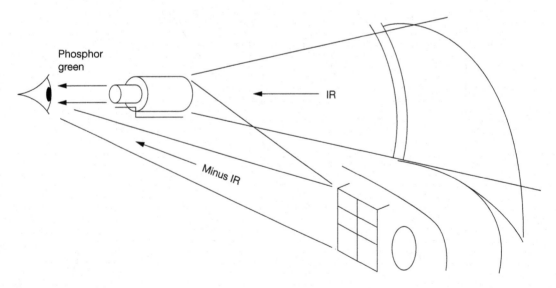

Phosphor green

IR

Minus IR

12.22 Minus IR filtering

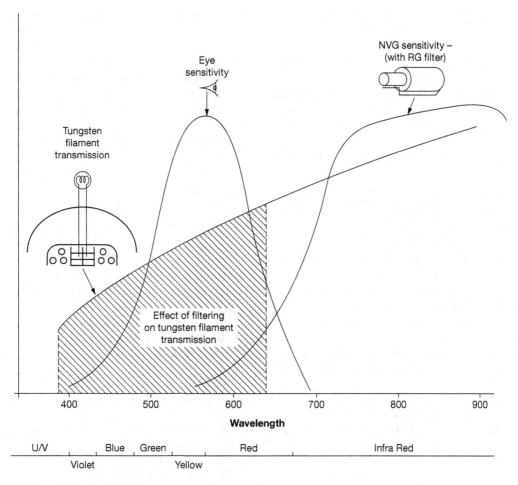

12.23 Tungsten filaments/filtering

MULTIPLE-CHOICE QUESTIONS

1. Bore-sighting is required on HUD components to ensure that:

   (a) The cockpit environment is NVG compatible

   (b) The desired chromaticity and IR filtering is achieved

   (c) Displayed images are referenced to the earth's horizon and the aircraft's projected flight path

2. Synthetic vision is based on:

   (a) Computer-generated displays that use three-dimensional (3D) images

   (b) On-board infrared (IR) sensors

   (c) Collimation technology that produces parallel light rays

3. Class B NVIS, is when the spectral content starts at:

   (a) 665nm, allowing wider colour use in avionics systems, with increased sensitivity
   (b) 665nm, allowing wider colour use in avionics systems, but with reduced sensitivity
   (c) 670nm, with a secondary peak at 540nm

4. The HUD flight-path vector (FPV) or velocity vector symbol indicates:

   (a) If the aircraft is accelerating or decelerating
   (b) The angle of attack
   (c) A predication of the aircraft's flight path

5. Near infrared light wavelengths is made of:

   (a) Visible wavelengths beyond 750nm
   (b) Invisible wavelengths beyond 750nm
   (c) Invisible wavelengths less than 750nm

6. Blooming refers to distorted NVG images surrounded with obscuring halos of light caused by:

   (a) Optical saturation of the goggles
   (b) The primary flying display (PFD)
   (c) Collimation technology that produces parallel light rays

7. The HUD energy-cue display indicates if the aircraft is:

   (a) On the desired track
   (b) Maintaining the desired glide path angle and touchdown point of the runway
   (c) Accelerating or decelerating

8. The spectral content of Class A NVIS starts at:

   (a) 625nm, enabling maximum use of ambient light
   (b) 665nm, enabling maximum use of ambient light
   (c) 625nm, enabling minimum use of ambient light

9. The spectral content above 700-750nm present in starlight is:

   (a) Visible to the unaided human eye
   (b) Not visible to the unaided human eye
   (c) Not visible with NVG

10. When flying in areas or at altitudes where rising terrain may pose a hazard, enhanced vision uses:

    (a) A terrain-alerting database to colourize the landscape
    (b) The primary flying display (PFD)
    (c) On-board sensors, e.g. FLIR

13 Fly-by-wire

Fly-by-wire (FBW) systems replace mechanical control components with electrical signals generated by position sensors, processed by a computer and then transmitted through electrical cables to the control actuators. FBW systems can be designed for fixed- and rotor-wing aircraft. Similar technology can also be employed to control the engines. A development of this technology is to replace the electrical cables with optical fibre, referred to as fly-by-light (FBL). FBW/FBL can be applied to both aeroplanes and rotorcraft.

Rather than mechanical linkages operating hydraulic actuators, fly-by-wire systems move flight-control surfaces (ailerons, rudders and so on) using electrical wire connections driving motors. At the heart of the system are computers that convert the pilot's commands into electrical signals which are transmitted to the motors, servos and actuators that drive the control surfaces. One problem with this system is the lack of 'feel' that the pilot experiences. Another is a concern over the reliability of FBW systems and the consequences of computer or electrical failure. Because of this, most FBW systems incorporate redundant computers as well as some kind of mechanical or hydraulic backup. Computer control reduces the burden on a pilot and makes it possible to introduce automatic control. Another significant advantage of FBW is a significant reduction in aircraft weight, which in turn reduces fuel consumption and helps to reduce undesirable CO_2 emissions.

HISTORY

As could be expected, the original research and development of FBW was conducted by the military. The first generation of commercial FBW was during the 1950s on the Sud Aviation Caravelle, a short/medium-range jet airliner; this featured electrical control of the yaw damper. Further developments on Concorde, VC10 and A300/310 aircraft saw the increasing use of electronic or electrically controlled flying surfaces, albeit with a mechanical back-up system. The first fully integrated full authority FBW system on a commercial passenger aircraft was introduced in 1988 on the Airbus A320. Recent examples of the next generation FBW are found in the A380 and B777 aircraft. Business/private aircraft, and smaller commercial air transport aircraft, including the Dassault Falcon 7X and Embraer 450/500, also employ FBW technology; these are described in more detail in the case studies in this chapter.

TEST YOUR UNDERSTANDING 13.1

Explain the difference between FBW and FBL.

Computer control can also help to ensure that an aircraft is flown more precisely and always within its 'flight protection envelope'. The crew are thus able to cope with emergency situations without running the risk of exceeding the flight envelope or overstressing the aircraft. Fly-by-wire technology has made it possible for aircraft manufacturers to develop 'families' of very similar aircraft. Airbus/EADS, for instance, has the 100-seat A318 to the 555-seat A380, with comparable flight-deck designs and handling characteristics. Crew training and conversion is therefore shorter, simpler and highly cost-effective. Additionally, pilots can remain current on more than one aircraft type simultaneously.

PRINCIPLES

Conventional flight control systems use a variety of cables and mechanical components driven directly by movement of the pilot controls. Cables run the entire length of the airframe from the cockpit controls to the various control surfaces – see Figure 13.1 for a simplified elevator control linkage. Cable-controlled

systems have high maintenance content and add weight to the aircraft. The mechanical system and its components need lubrication and rigging adjustments to compensate for long-term cable stretching. On a larger commercial transport aircraft, there are many more control surfaces (see Figure 13.2); the weight and maintenance costs for a mechanical system becomes prohibitive.

In an FBW flight-control system, the control of flight-control surfaces is via actuators that are controlled electrically – see Figure 13.3 for a simplified system. FBW computers convert electronic signals from position transducers attached to the pilot controls; these input signals are processed and converted into output commands that are transmitted to the flying-control surface actuators.

Aside from reducing the maintenance requirements for cables and mechanical components, FBW also offers additional features and benefits:

- Improved flight handling
- Integration of flight-control functions
- Reduced size of flight-control surfaces
- Envelope protection
- Envelope alerting

By thorough study, analysis and understanding of the aircraft's flight characteristics, the control laws and system functionality can be optimized to give

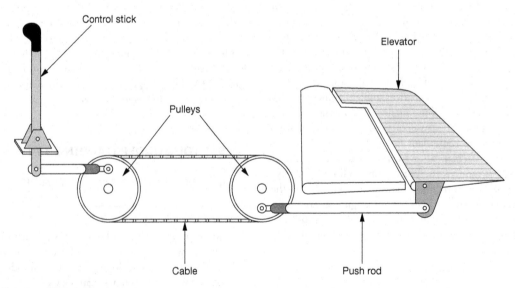

13.1 Simplified elevator control

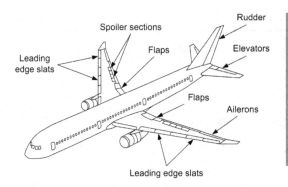

13.2 Large aircraft control surfaces

KEY POINT

FBW computers convert electronic signals from position transducers attached to the pilot controls; these input signals are processed and converted into output commands that are transmitted to the flying-control surface actuators.

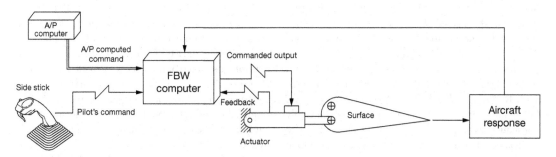

13.3 FBW comparison

improved handling. This is achieved by retaining the essential characteristics of a conventionally controlled system, and eliminating (or reducing) undesirable characteristics. Coordination of control inputs to multiple surfaces requires high skill levels and concentration from the pilot; automation of these control functions in response to pilot commands reduces the need for other stability-augmentation systems and components, e.g. the rudder can be controlled by a single channel of the control system for both turn coordination and yaw damping, thereby deleting a dedicated yaw damper system. Advanced control laws give much faster reaction times compared to what a pilot can achieve. The size of the flight-control surfaces can be made smaller, resulting in an overall reduction in system weight.

Envelope protection dynamically controls the flight of the aircraft in order to prevent stalls or over-speed situations that might otherwise be induced by the autopilot or flight director commands. This feature continuously accounts for all sources of lift demand (pitch, bank, altitude, speed, weight), computes safe limits on the flight condition and constrains the

autopilot to fly within these safe limits. Envelope alerting dynamically provides the pilot with full-time audible and visual alerting in the event that aircraft is flown outside of its safe limits and even when the autopilot and flight director are not actively engaged.

Although there are benefits with FBW, this comes at cost during the initial design and certification of new systems. The system has to have very stringent dependability requirements both in terms of:

• Safety: the system must not generate non-commanded signals outputs
• Availability: the complete loss of the system must be extremely improbable

The overall safety and availability of the FBW system is based on the system architecture, tolerance to both hardware and software failures and failure monitoring. These lead to high non-recurring costs during the development of a new aircraft type or variant. System architecture is typically based on multiple computers that have dissimilar software to reduce the risk of erroneous signal outputs. Redundancy is built into the

Table 13.1 Software criticality (CS25 large aeroplanes)

Level	Classification	Effect	Qualitative	Quantitative
A	Catastrophic	Normally loss of aircraft	Extremely improbable	$<10^{-9}$
B	Hazardous	Large reduction in safety	Extremely remote	$<10^{-7}$
C	Major	Significant reduction in safety	Remote	$<10^{-5}$
D	Minor	Slight reduction in safety	Probable	$<10^{-3}$
E	None	No safety effect	None	None

system, such that a single computer can be disconnected from the control system, without affecting control of the aircraft. This disconnection can either be done by the computer's own self-testing or via cross-checks from other computers.

The software used in FBW computers is validated to the highest specification (Level A) in accordance with RTCA D0178/ ED-12. Aircraft software can be divided into five levels according to the likely consequences of its failure, as shown in Table 13.1. The highest level of criticality (Level A) is that which would have catastrophic consequences, whilst the lowest level of criticality is that which would have no significant impact on the operation of the aircraft. In between these levels the degree of criticality is expressed in terms of the additional workload imposed on the flight crew and, in particular, the ability of the flight crew to manage the aircraft without having access to the automatic control or flight information that would have otherwise been provided by the failed software.

The initial certification of an aircraft requires that the design organization (DO) shall provide evidence that the software has been designed, tested and integrated with the associated hardware in a manner that satisfies standard DO-178/ED-12 (or an agreed equivalent standard). In order to provide an effective means of software identification and change control, a software configuration management plan (CMP) is required to be effective throughout the life of the equipment (the CMP must be devised and maintained by the relevant DO).

Post-certificate modification of equipment in the catastrophic, hazardous or major categories (Levels A, B and C) must not be made unless first approved by the DO. Hence all software upgrades and modifications are subject to the same approval procedures as are applied to hardware modifications. This is an important point that recognizes the importance of software as an 'aircraft part'. Any modifications made to software must be identified and controlled in accordance with the CMP.

KEY POINT

The software used in FBW computers is validated to the highest specification (Level A).

The software level, also known as the design-assurance level (DAL), is determined from the safety-assessment process and hazard analysis by examining the failure effects of the system. The safety-assessment process is a systematic review of each functional aspect of the system, how it can go wrong, what happens when it goes wrong and the mitigation.

SYSTEM OVERVIEW

Two of the world's largest aircraft manufacturers are Boeing and Airbus, and each have different philosophies with respect to FBW. The Boeing 777 retains the conventional control column, control wheel and rudder pedals; operation of these controls is identical to other Boeing commercial transport aircraft. The Airbus design is based on a side-stick controller, offering reduced size and weight, together with increased visibility of the instrument panel.

Whichever design is used, the pilot's control incorporates transducers that convert physical displacement into electrical signals. To avoid any over-control in the conventional system, the tactile feel of the control input is sensed by proportionally increasing the amount of force the pilot experiences during a manoeuvre; this is matched to the feel that would exist with a conventional control system.

Control surfaces on the wing and tail of a typical system are controlled by hydraulically powered, electrically signalled actuators. Elevators and ailerons are controlled by two actuators per surface, and the rudder is controlled by three. Each spoiler is powered by a single actuator. The horizontal stabilizer is positioned by two parallel hydraulic motors driving the stabilizer jackscrew.

The current generation of FBW systems have emergency reversion, or back-up, using mechanical linkages. These are only for continued safe flight and landing, rather than full authority control and protection.

CASE STUDIES

Two business aircraft that have been designed with FBW controls are the Embraer 450/500 and Dassault Falcon 7X. A summary of each system's architecture follows.

Embraer 450/500 FBW

The main components of the Embraer FBW flight-control architecture are illustrated in Figure 13.4. The primary features include cockpit inceptors (pilot and co-pilot side sticks, rudder pedals, speed brake, flap handle), inceptor interface modules (IIM), remote electronic units (REU) and the primary flight-control computer (FCC). The IIMs are responsible for reading cockpit inceptor sensors and sending the data to the FCC through the digital data bus. The FCCs provide the main computing functions; there are two redundant FCCs, each of which has dual dissimilar lanes, in a command/monitor scheme. The two FCC channels operate in an active/standby configuration. The REU is the electronic unit that operates the airplane's flight-control surface actuators. The REUs operate the primary control surfaces, and also the spoilers, flaps and horizontal stabilizer.

Cockpit side sticks and pedals

Embraer is using side-stick controls for the Legacy 450/500, instead of the conventional control wheel and yoke system. This offers a better view of the displays, a cleaner cockpit and more precise control of the airplane, as well as reduced weight, maintenance and spare parts.

The side stick is passive, where force feedback is provided by dual fixed springs, each tuned to the longitudinal and lateral axis respectively. If dual pilot input is computed, aural and visual warnings and stick vibration alert that both pilots are making inputs. The rudder pedals are mechanically linked with dual fixed springs, also providing a feel of the force of the airplane's movement.

Flight-control laws

For the Legacy 450 and 500, the flight-control computers (FCC) compute the commands to move the flight-control surfaces based on input from the pilots via the side stick and the rudder pedals. In normal operations (that is, without significant system failures), the pilot commands are interpreted by the FCCs as demands for aircraft response rather than surface deflection, as in a conventional aircraft. The FCCs move the surface, in order to achieve the desired response.

In normal operations, the longitudinal control law maintains the aircraft flight path and stick deflection command flight path rate. This means that the Legacy 450 and 500 do not behave like a conventional aircraft; they are flight-path stable, rather than speed stable. A conventional aircraft will maintain its trimmed speed. If velocity is lost, the aircraft pitches down to maintain the trimmed airspeed and, likewise, if airspeed increases, the aircraft will pitch up to maintain velocity. Thus the conventional airplane is speed stable, rather than flight-path stable.

The Legacy 500 control law for most of the flight phases is flight-path stable. When the aircraft begins to change speed the control law maintains the flight path. The control law automatically trims the aircraft to the new speed. This significantly reduces pilot workload and is totally compatible with side-stick operation, because the stick can stay in the neutral position while the control law trims automatically.

The control law provides conventional speed stability during the landing-flare flight phase. When

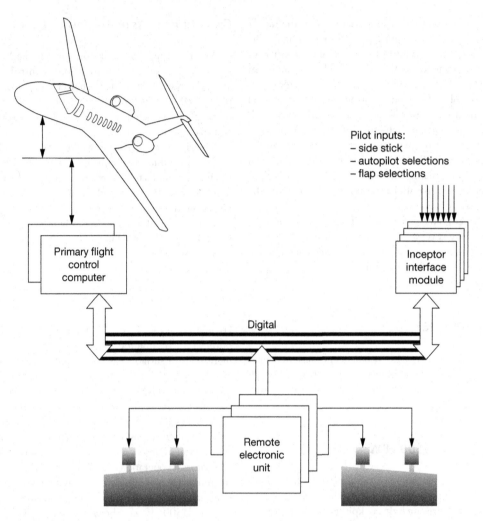

13.4 Embraer FBW architecture

the aircraft begins to lose speed, the pilot has to pull back on the stick, in order to prevent a pitch-down movement, like with a conventional aircraft.

During turns, the control law will automatically compensate for the pitch command required to maintain the flight path to a bank angle of up to 33 degrees. There is no side-stick back pressure required to maintain the flight path for bank angles up to 33 degrees.

The lateral flight-control law for the Legacy 500 maintains the given bank-angle rate based on side-stick deflection and provides neutral spiral stability for bank angles of up to 33 degrees, and strong positive spiral stability beyond 33 degrees. Thus, the control law will indirectly maintain the given bank angle at 33 degrees, with the side stick at neutral, and roll the aircraft back to 33 degrees, if the side stick is in the neutral position and the bank angle is greater than 33 degrees. Combined with automatic pitch compensation of a bank angle up to 33 degrees, this allows safe recovery from roll upset by returning the stick to the neutral position.

The lateral control law also significantly reduces pilot workload. The Legacy 500 can automatically keep a steady rate of turn with the side stick in the neutral position. During situations when it is necessary to maintain a side-slip angle (e.g. when one engine is inoperative), the control law significantly reduces workload, because it provides most of the compensation automatically, leaving only a residual for the pilot to recognize engine failure.

When it is not possible to maintain the normal mode operation, due to multiple system failures, the flight controls revert to direct-mode operation. In this case, the normal mode-control laws are replaced by direct-mode control laws. The inceptor commands go directly to the REUs that have a gain schedule based only on air data or, in the case of loss of air data, flap position information. During this operation, the aircraft behaves like a conventional aircraft, and the side stick and rudder pedal deflection are directly related to the surface positions.

Flight envelope and protection

One of the key aspects of the Legacy 450 and 500 fly-by-wire controls is flight-envelope protection. The control laws prevent the aircraft from exceeding the limitation of the flight envelope, such as structural load, high- and low-speed limits, and stall.

The flight-envelope protections for the Legacy 450 and 500 are used for structural-load factor (G protection), angle of attack (stall/buffeting protection), dive speed (high-speed protection) and side-slip (lateral structural-load protection). The control laws calculate the appropriate surface commands to keep the aircraft inside the limits. Hard flight-envelope protection is not applied to bank and pitch angle limits. If necessary, the pilot can make aggressive manoeuvres to exceed these limits.

A concept applied to the project in the design of the flight-control laws is the definition of two envelopes for the aircraft. One is the normal flight envelope; inside this envelope, the control of the aircraft requires less pilot workload, and provides significant passenger comfort. The control law will keep the aircraft trimmed to any condition inside this envelope, with the stick in the neutral position. The limits of the normal flight envelope are a bank angle of 33 degrees, maximum operational speed, 1.1 stall speed, +30 or −15 degrees of pitch angle.

Beside the normal flight envelope, there is the flight envelope limit. This envelope is when the aircraft is close to the extreme envelope limits, and a greater pilot workload is needed to maintain the aircraft in this condition. The pilot cannot override the extreme limits and, if the stick is released, the aircraft is automatically brought back to the normal flight envelope. The hard limits of this envelope are the aircraft structural limits, maximum angle of attack, dive speed and side-slip.

Falcon 7X FBW

Dassault elected to fit the Falcon 7X with one of the most redundant FBW systems ever installed in a civil aircraft. It is a highly fault-tolerant system, with extensive cockpit integration (see Figure 13.5). Each pilot has a side-stick control input, as illustrated in Figure 13.6.

Dispatch is permitted with multiple single-component faults, including failure of one channel of

13.5 Dassault Falcon 7X cockpit

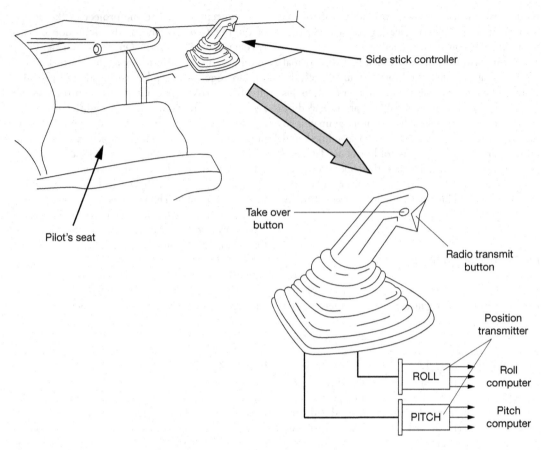

13.6 Side stick

each side-stick controller, one SmartProbe® rendered inoperative and the loss of a single flight-data concentrator, one channel in a main flight-control computer (MFCC) or in an actuator control-monitoring unit. System architecture is illustrated in Figure 13.7.

FBW systems are critically dependent on electrical power, so three engine-driven generators, two engine-driven permanent magnet alternators and a ram air turbine (RAT) generator are available for power supply. Two batteries can power the system, if all other power sources fail. The auxiliary power unit (APU) only supplies electrical power on the ground. The Falcon 7X's FBW system may be divided into five areas, as illustrated in Figure 13.8:

1. Sensors
2. Data concentrators
3. Flight-control computers
4. Actuator controllers
5. Actuators

Pilot control and sensor inputs are fed to the flight-data concentrators. The data-acquisition units supply three dual-channel main flight-control computers (MFCCs) and three single-channel secondary flight-control computers (SFCCs). The location of these computers is split between forward and aft ends, and left and right sides of the fuselage, for maximum isolation. The routing of FBW wiring harnesses is widely separated for optimum damage tolerance.

The first level of the FBW system is comprised of pilot controls and sensors, the intelligence gatherers of the system. Each group of sensors supplying the FBW system has multiple channels. Most individual boxes are at least dual-channel designs. Each SmartProbe®, for instance, is dual-channel. Each side-

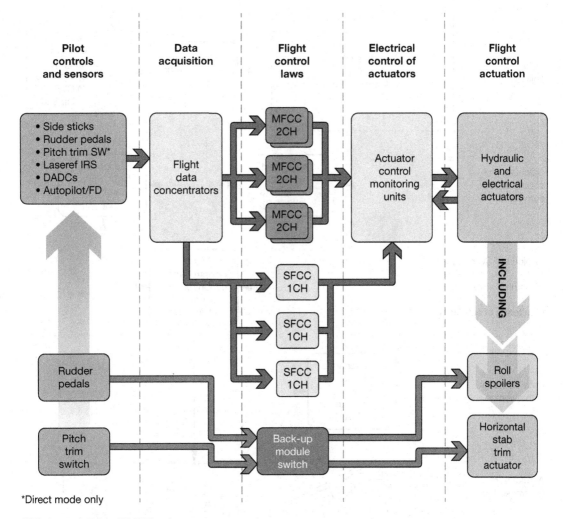

*Direct mode only

13.7 Dassault Falcon 7X FBW architecture

stick controller is a five-channel unit because of its critical functionality. The redundancy is so complete that operators will be able to dispatch with a single failed channel in multiple groups of sensors and still have adequate backup in the case of a subsequent channel or line replaceable unit (LRU) failure.

The second level of the system is dedicated to data concentrating, processing all the inputs for use by the flight-control computers. The input sensors feed analogue, digital and discrete signals to five dual-channel flight-data concentrators, although there is only one LRU depicted on the accompanying diagram. The data concentrators digitize all the inputs and forward them to the flight-control computers.

The third level of the system is the flight-control computer section. Normally, the three dual-channel main flight-control computers handle all aspects of control functions; each channel in the MFCCs runs different software. Each channel either generates flight-control laws or monitors the results, with roles alternating at each power up. MFCCs are capable of normal, alternate and direct law modes. Any one of the three MFCCs can provide complete FBW normal-law-mode functionality. All of the stability enhancements, envelope protection functions, auto trim and auto flight control configuration features remain available in the normal-law mode so long as redundant input, sensor and interface inputs are

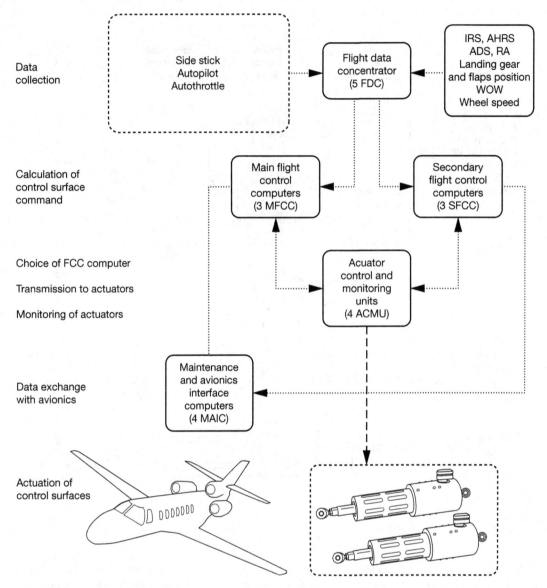

13.8 Dassault Falcon 7X FBW overview

available. Normal law, therefore, needs a 'second opinion' from every input source to verify the integrity of the primary input, plus a 'second opinion' from the alternate channel of at least one MFCC. If there is only one signal from any group of input sources then the system automatically degrades to the alternate law mode. Some or all of the normal-law functions may be lost, depending upon which input source has been lost.

The MFCCs' direct-law mode is only invoked if all signals from certain critical sensors are lost. At that point, the pilot is flying the aircraft with simple electrical links to the flight controls that emulate the mechanical linkages of a conventional flight control system. The Falcon 7X has relaxed static and dynamic stability characteristics in direct-law mode; there is a 10^{-7} probability of this level of degradation occurring. The single-channel SFCCs are only capable of operating in the direct-law mode; that is because they lack a second channel needed to verify the integrity of the decision-making process of the primary channel. However, the SFCCs have different microprocessors than the MFCCs and they run completely different software. Their function is to provide simple but triple-redundant critical flight-control functionality in the event that all three MFCCs fail.

Control-law signals from the MFCCs or SFCCs are sent to the actuator control and monitoring units. The actuator control and monitoring units (ACMUs) command the positioning of the primary and secondary flight-control actuators, in essence replacing the mechanical links to the servo actuators that would be found in a conventional flight-control system. Position data from the actuators is fed back to the ACMUs to verify that the commanded position occurred. The ACMUs also perform reasonability checks on signals received from the main and secondary flight-control computers.

Electric and electro-hydraulic actuators are used to move control surfaces; three hydraulic systems supply the actuators. The aircraft is controllable with two of the three systems inoperative. The basic FBW system architecture has a 10^{-9} probability of failure, thus meeting the most stringent certification requirements. The Falcon 7X is fitted with an emergency backup flight-control system that uses discrete inputs from the rudder pedals and pitch trim switch. These inputs are separate from the redundant channels that feed the normal FBW system and they bypass all other second-, third- and fourth-level components. The backup module electrically trims the position of the horizontal stabilizer for pitch control. Inputs from the rudder pedals command the position of the roll spoilers, not the rudder! The roll spoilers use C system hydraulic power supplied by the number two engine-driven pump. But if the engine-driven pump is inoperative, electrically powered backup hydraulic actuators in the wings power the roll spoilers. The intent of the backup system is to provide the crew with additional time to reset or recycle the MFCCs and SFCCs, thereby restoring a higher level of FBW control.

ROTORCRAFT

Fly-by-wire technology applications in rotorcraft present additional requirements due to unique safety and reliability considerations. The primary concern is loss of rotor pitch control, possibly leading to loss of the rotorcraft. In general terms, the use of FBW in production rotorcraft has lagged behind that in fixed-wing aircraft in the civil field. That said, fly-by-wire technology is in production or development on civil applications, including the Sikorsky S92. As rotorcraft designs have been developed and matured, their speeds, manoeuvrability and performance capabilities have increased, subsequently making them more difficult to fly without stability augmentation (as described in an earlier chapter) and other electro-mechanical systems.

Control commands from the pilot and the rotorcraft sensors – e.g. air data, attitude, accelerometers and so on – are fed to a flight-control computer (FCC) as electrical input signals. The FCC then integrates these various inputs and generates output commands to the various flying-control actuators. Rotorcraft FBW offers the following:

• Removal of mechanical linkages
• Weight savings
• Reduced maintenance costs
• Reduced pilot workload
• Increased handling capability and manoeuvrability

Referring to Figure 13.9, the side-arm controller (SAC) is made achievable through FBW/FBL; it allows either pilot to operate the controls with one hand. The SAC is mounted to the side of each pilot, allowing increased visibility of the instrument panel. It operates in the conventional sense for cyclic pitch and roll, and is twisted for yaw control. The SAC is moved vertically for collective control.

KEY POINT

Envelope protection dynamically controls the flight of the aircraft in order to prevent stalls or over-speed situations.

The primary consideration for rotorcraft FBW is the pilot-induced oscillation (PIO). This is an inadvertent, sustained oscillation of the rotorcraft arising from an abnormal handling condition. PIO involves sustained or uncontrollable oscillations resulting from the pilot

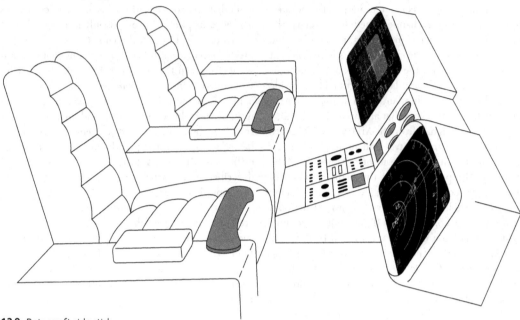

13.9 Rotorcraft side stick

trying to control the rotorcraft. Digital FBW control systems offer the potential for managing the interactions between the pilot and rotorcraft dynamics. Rotorcraft have the inherent dynamics resulting from the interactions between the main and tail rotors, and the airframe. These dynamics include translational flight, pendulous motion and rotor gyroscopic effects, as described in an earlier chapter. The dominant variables are rotor speed and torque loads that result from the rotorcraft motion. Furthermore, rotorcraft FBW has to consider:

- Integration of flight and propulsion control
- Autorotation and hover
- Low altitude operation
- External loads, cargo etc.

KEY POINT

FBW system architecture is typically based on multiple computers that have dissimilar software.

KEY POINT

Rotorcraft pilot-induced oscillation (PIO) is an inadvertent, sustained oscillation of the rotorcraft arising from an abnormal handling condition.

FADEC

Engines can also be controlled electronically, rather than by mechanical linkages; the full-authority digital engine control (FADEC) comprises a digital computer, the electronic engine controller (EEC) or engine-control unit (ECU), and its related components. The system controls all aspects of an aircraft engine's performance; FADECs are applicable to both piston and jet engines, for both aeroplanes and rotorcraft. FADEC technology gives optimum engine efficiency for a given flight condition, and also controls engine starting and restarting.

The FADEC replaces the mechanical linkages, cables, rods and so on between the throttle levers and the fuel-control unit (FCU) that controls the engine's speed and thrust. The FCU itself is a hydro-mechanical governor that measures various engine sensors, e.g. RPM, pressures and temperatures, to control engine thrust to the power setting selected by the pilot. The FCU also compensates for atmospheric temperature, barometric pressure and altitude changes to maintain the selected engine speed or thrust. (A full account of engine sensors is given in another title in this book series, *Aircraft Electrical and Electronic Systems*.)

A typical gas turbine engine FADEC computer is located in the airframe, or on the fan casing, and is typically designed with dual-channel architecture; some systems are based on triple-redundant system architecture. Each channel, or lane, comprises a control and monitoring function, as seen in Figure 13.10. The thrust demand from the throttle lever is processed in accordance with control laws to manage the flow of fuel into the engine.

In a piston engine, the typical inputs are from:

- Shaft speed
- Cylinder-head temperature
- Exhaust-gas temperature
- Manifold air pressure and temperature
- Fuel-pressure sensors
- Throttle-position switch

The FADEC continuously monitors and controls ignition timing, fuel injection timing and fuel-to-air-ratio mixture, thereby eliminating the need for magnetos and manual fuel/air-mixture control.

MULTIPLE-CHOICE QUESTIONS

1. The software used in FBW computers is validated to:

   (a) Level B in accordance with RTCA D0178/ ED-12
   (b) Level A in accordance with RTCA D0178/ ED-12
   (c) Level A in accordance with RTCA D0160

2. Envelope protection:

   (a) Dynamically controls the flight of the aircraft in order to prevent stalls or over-speed situations
   (b) Dynamically provides the pilot with full-time audible and visual alerting in the event that aircraft is flown outside of its safe limits
   (c) Is an inadvertent, sustained oscillation of the rotorcraft arising from an abnormal handling condition

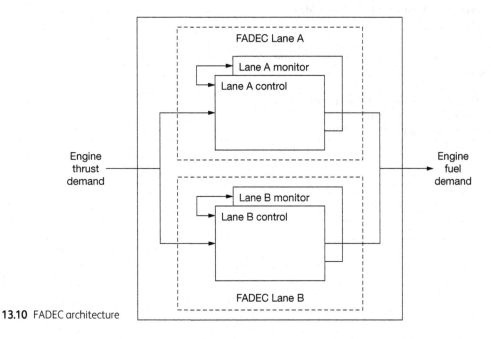

13.10 FADEC architecture

3. Envelope alerting:

   (a) Dynamically controls the flight of the aircraft in order to prevent stalls or over-speed situations
   (b) Prevents sustained oscillation of the rotorcraft arising from a normal handling condition
   (c) Dynamically provides the pilot with full-time audible and visual alerting in the event that the aircraft is flown outside of its safe limits

4. FBW system architecture is typically based on:

   (a) Multiple computers that have identical software
   (b) Multiple computers that have dissimilar software
   (c) Single-channel computers

5. Rotorcraft pilot-induced oscillation (PIO) is:

   (a) An inadvertent, sustained oscillation of the rotorcraft arising from an abnormal handling condition
   (b) A sustained oscillation of the rotorcraft arising from a normal handling condition
   (c) A full-time audible and visual alerting in the event that the aircraft is flown outside of its safe limits

6. The rotorcraft SAC operates in the conventional sense for:

   (a) Moved vertically for pitch and roll
   (b) Twisted for cyclic pitch and roll
   (c) Cyclic pitch and roll; and is twisted for yaw control

7. Emergency FBW reversion using mechanical linkages is only for:

   (a) Continued safe flight and landing
   (b) Full-authority control and protection
   (c) Pilot-induced oscillation

8. Failure of Level A software in certification terms would have:

   (a) Catastrophic consequences
   (b) No safety effect
   (c) Hazardous consequences

9. The probability of complete loss of the FBW system must be

   (a) Remote
   (b) Extremely improbable
   (c) Extremely remote

10. In a FADEC aircraft, the thrust demand from the throttle lever is processed in accordance with control laws to manage the:

    (a) Shaft speed
    (b) Turbine temperature
    (c) Flow of fuel into the engine

Appendix A: Glossary

This appendix gives an alphabetical list of terms in this book.

A/T	autothrottle
ACMU	actuator control and monitoring unit
ACNS	aircraft communications and navigation systems
ACP	autopilot control panel
ACU	analogue converter unit
ADAHRS	air data, attitude and heading reference system
ADC	air data computer
ADF	automatic direction finder
ADI	attitude direction indicator
AEES	aircraft electrical and electronic systems
AEP	aircraft engineering principles
AFCS	automatic flight control system
AHRS	attitude heading reference system
AHS	attitude hold system
AI	attitude indicator
ALA	approach and landing accidents
ALT	altitude
AM	amplitude modulated
ANVIS	aviator's night vision imaging system
AOA	angle of attack
AOB	angle of bank
AOI	angle of incidence
AOU	auxiliary power unit
AP	autopilot
APV	approach with vertical guidance
AR	(A)NVIS radiance
ASA	autoland status annunciator
ASE	automatic stabilization equipment
ASI	air-speed indicator
ATC	air traffic control
BIT	built-in test
BLEU	Blind Landing Experimental Unit
C of G	centre of gravity
CADC	central air data computer
CAN	controller area network
CAS	calibrated airspeed
C_D	coefficient of drag
CDI	course deviation indicator
CDU	control display unit
CFIT	controlled flight into terrain
CG	centre of gravity
CIE	International Commission on Illumination
C_L	coefficient of lift
CM	configuration module
CMP	configuration management plan
CNC	computer numerical control
CO_2	carbon dioxide
COTS	commercial off the shelf
CP	centre of pressure
CPU	central processing unit
CRT	cathode ray tube
CS	certification specification
CTD	cross track deviation
CVS	combined vision systems
DA	drift angle
DAL	design assurance level
DFC90/100	Avidyne autopilot
DFGC	digital flight guidance computer
DG	directional gyro

DH	decision height	GS	ground speed
DIS	distance	HDD	head-down display
DME	distance-measuring equipment	HDG	heading
DO	design organization	HFE	human factors engineering
DP	departure procedures	HSI	horizontal situation indicator
DSRTK	desired track angle	HUD	head-up display
EADI	electronic ADI	I^2T	image-intensifying tube
EADS	European Aeronautic Defence and Space Company	IAS	indicated air speed
		ICAO	International Civil Aviation Organization
EAS	equivalent air speed	IFR	instrument flight rules
EASA	European Aviation Safety Agency	IIM	inceptor interface module
ECU	engine control unit	IIT	image-intensifying tube
EEC	electronic engine controller	ILS	instrument landing system
EEPROM	electronically erasable programmable read only memory	INS	inertial navigation system
		IOP	input-output processor
EFB	electronic flight bag	IR	infrared
EFD	electronic flight display	IRS	inertial reference system
EFD	Evolution Flight Display (Aspen Avionics)	ISA	International Standard Atmosphere
		L/D	lift-to-drag ratio
EFIS	electronic flight instrument system	LAAS	local area augmentation system
EGNOS	European geostationary navigation overlay service	LBS	lateral beam sensor
		LCD	liquid crystal display
EP	Envelope Protection (Avidyne EP™)	LDI	lateral deviation indicator
EPR	engine pressure ratio	LED	light-emitting diode
ESA	European Space Agency	LF	low-frequency
ETL	effective translational lift	LOC	localizer
EVS	enhanced vision system	LPV	localizer performance with vertical guidance
FADEC	full authority digital engine control		
FAR	federal aviation requirement	LRRA	low-range radio altimeter
FBL	fly-by-light	LRU	line replaceable unit
FBW	fly-by-wire	LVDT	linear variable displacement transducer
FCC	flight-control computer	MAP	main application processor
FCC	fuel-control unit	MCP	mode control panel
FD	flight director	MEL	minimum equipment list
FDE	fault detection and exclusion	MEMS	micro-electrical-mechanical sensors
FLIR	forward-looking infrared	MF	medium-frequency
FMA	flight-mode annunciator	MFCC	main flight control computer
FMCW	frequency modulated, continuous wave	MFD	multi-function display
FMS	flight management system	MLS	microwave landing system
FOG	fibre-optic gyro	MMWR	millimetre wave radar
FOQA	flight operations quality assurance	MSP	mode-select panel
FPM	feet per minute	MTBF	mean time between failure
FPV f	light path vector	NAA	national aviation authority
GA	general aviation	NM	nautical mile
GA	go-around	NPA	non-precision approach
GLS	GPS landing system	NRa	NVIS radiance for class A
GNSS	(European) global navigation satellite system	NRb	NVIS radiance for class B
		NRc	NVIS radiance for class C
GPS	global positioning system	NVG	night-vision goggles
GS	glide slope	NVIS	night-vision imaging system

OAT	outside air temperature	SSR	secondary surveillance radar
OEM	original equipment manufacturer	STAR	standard terminal arrival route
PA	precision approach	SVS	synthetic vision system
PC	pitch computer	SVT™	synthetic vision technology (Garmin)
PCB	printed circuit board	TA	transition altitude
PFD	primary flying display	TAE	track angle error
PIO	pilot-induced oscillation	TAS	true airspeed
POS	present position	TAWS	terrain awareness and warning system
PVD	para-visual display	TCAS	traffic collision avoidance system
RAIM	receiver autonomous integrity monitoring	TIS	traffic information services
		TK	ground track angle
RAT	ram air turbine	TKE	track angle error
REU	remote electronic unit	TOGA	takeoff/go-around
RFI	radio frequency interference	VBS	vertical beam sensor
RLG	ring laser gyros	VDI	vertical deviation indicator
RNAV	area navigation system	VDL	VHF data links
ROC	rate of climb	VFR	visual flight rules
RPM	revolutions per minute	VG	vertical gyro
RPS	regional forecast pressure	VHF	very high frequency
RSM	remote sensor module	VNAV	vertical navigation
RTCA	Radio Technical Commission for Aeronautics	VOR	VHF omnirange
		VS	vertical speed
RVDT	rotary variable displacement transducer	VSI	vertical speed indicator
		W	weight
RVR	runway visual range	WAAS	wide area augmentation system
RVSM	reduced vertical separation minimums	WD	wind direction
SAC	side arm controller	WS	wind speed
SAS	stability augmentation system	XTK	cross track distance
SCAS	stability control and augmentation system		

Appendix B: Revision questions

The following 120 questions are based on subjects covered throughout this book. Attempt all questions and aim to achieve over 100 correct answers.

1. When carrying out a pressure-leak test on an altimeter, this will check the:

   (a) Capsules
   (b) Instrument case
   (c) Accuracy

2. ASIs are supplied with which pressure inputs?

   (a) Dynamic
   (b) Static
   (c) Dynamic and static

3. On descent, the pressure inside a VSI capsule:

   (a) Is less than the case pressure
   (b) Is higher than the case pressure
   (c) Equals the case pressure

4. Position error is caused by the:

   (a) Design and construction of the unit
   (b) ASI being calibrated for sea-level air density
   (c) Sensed static pressure not being the true ambient still-air pressure

5. If a static drain trap is removed, but no water is found, the system must be:

   (a) Purged
   (b) Leak tested
   (c) Calibrated

6. Instrument accuracy is used to determine:

   (a) Truth and comparison with a true value
   (b) Exactness and repeatability
   (c) The difference between the true reading and the observed reading

7. Dead reckoning is the process of:

   (a) Correcting the aircraft's position
   (b) Estimating the aircraft's position
   (c) Fixing the aircraft's position

8. The angle between the aircraft's heading and ground track is known as the:

   (a) Drift angle
   (b) Cross track distance
   (c) Wind vector

9. Magnetic compasses are unreliable in the:

   (a) Short term
   (b) Equatorial regions
   (c) Long term, flying a constant heading

10. The angle between north and the flight path of the aircraft is the:

    (a) Ground track angle
    (b) Drift angle
    (c) Heading.

11. Temperature decreases uniformly with altitude until about:

    (a) 36,000 feet (11km)
    (b) 65,000 feet (20km)
    (c) 105,000 feet (32 km)

12. The ease with which a fluid flows is an indication of its:

    (a) Humidity
    (b) Temperature
    (c) Viscosity

13. Temperature in the upper stratosphere starts to:

    (a) Decrease after 65,000 feet
    (b) Increase after 65,000 feet
    (c) Increase after 36,000 feet

14. For an ideal fluid, the total energy in a steady streamline flow:

    (a) Remains constant
    (b) Increases
    (c) Decreases

15. An average value for the tropopause in the International Standard Atmosphere is around:

    (a) 105,000 feet (32km)
    (b) 65,800 feet (20km)
    (c) 36,000 feet (11km)

16. The amount of water vapour that a gas can absorb:

    (a) Decreases with increase in temperature
    (b) Decreases with decrease in temperature
    (c) Increases with decrease in temperature

17. The three layers of atmosphere nearest the surface of the earth are known as the:

    (a) Stratosphere, chemosphere, troposphere
    (b) Chemosphere, troposphere, stratosphere
    (c) Troposphere, stratosphere, chemosphere

18. With increasing altitude, the atmospheric pressure decrease will be:

    (a) Non-linear
    (b) Linear
    (c) Highest in the stratosphere

19. The boundary between the troposphere and stratosphere is known as the:

    (a) Tropopause
    (b) Stratopause
    (c) Ionosphere

20. Compared with dry air, humid air is:

    (a) Heavier, or less dense
    (b) Lighter, or less dense
    (c) Lighter, or more dense

21. The direction in which precession takes place in a gyroscope is dependent upon the:

    (a) Rotor mass and speed
    (b) Direction of rotation and the axis about which the torque is applied
    (c) Location on the earth's surface

22. The attitude indicator displays:

    (a) Pitch and roll
    (b) Heading
    (c) Rate of climb

23. The turn coordinator is often used in place of the:

    (a) Artificial horizon
    (b) Attitude indicator
    (c) Turn and slip indicator

24. Low gyro wheel speeds cause:

    (a) Slow instrument response or lagging indi-
        cations
    (b) The instrument to overreact
    (c) No effect on the indication

25. Turn indicators are based on which type of
    gyroscope?

    (a) Displacement
    (b) Rate
    (c) Vertical

26. The turn coordinator displays:

    (a) Pitch attitude
    (b) Roll attitude
    (c) Predefined roll rate

27. The rate gyro responds to:

    (a) Changes in direction
    (b) Displacement
    (c) Magnitude of direction

28 The directional gyro is used to sense direction
    in the:

    (a) Azimuth plane, with a horizontal spin axis
    (b) Azimuth plane, with a vertical spin axis
    (c) Longitudinal plane, with a horizontal spin
        axis

29. Laser gyros sense:

    (a) Heading
    (b) Angular rate of rotation about an axis
    (c) Displacement

30. Rigidity of a gyroscope increases with:

    (a) Applied torque
    (b) Location on the earth's surface
    (c) Rotor mass and speed

31. An open-loop control system is one in which:

    (a) No feedback is applied
    (b) Positive feedback is applied
    (c) Negative feedback is applied

32. The range of outputs close to the zero point
    that a control system is unable to respond to is
    referred to as:

    (a) Hunting
    (b) Overshoot
    (c) Deadband

33. A closed-loop control system is one in which:

    (a) No feedback is applied
    (b) Negative feedback is applied
    (c) Positive feedback is applied

34. The gain of a servo control system determines:

    (a) How fast the servo motor tries to reduce
        the error
    (b) Damping
    (c) If it is open- or closed-loop

35. Servo system feedback has to be negative to
    minimize:

    (a) Deadband
    (b) Gain
    (c) Oscillations

36. Overshoot in a control system can be reduced
    by:

    (a) Reducing the damping
    (b) Increasing the gain
    (c) Increasing the damping

37. Decreasing the gain of a servo system will:

    (a) Reduce the overshoot and slow down the
        response
    (b) Reduce the overshoot and speed up the
        response
    (c) Increase the overshoot but slow down the
        response

38. The optimum value of damping in a control
    system is that which allows:

    (a) No overshooting
    (b) Increasing overshooting
    (c) One small overshoot

39. Ramp inputs in a control system are characterized by:

    (a) Gradually changing at a given speed
    (b) Changing instantaneously
    (c) High-gain servo systems

40. The output of a control system continuously oscillating above and below the required value is known as:

    (a) Deadband
    (b) Hunting
    (c) Overshoot

41. The angle of attack is the angle between the:

    (a) Chord line and the relative airflow
    (b) Relative airflow and the longitudinal axis of the aircraft
    (c) Maximum camber line and the relative airflow

42. If the angle of attack of an aerofoil is increased, the centre of pressure will:

    (a) Move backward
    (b) Stay the same
    (c) Move forward

43. The angle of incidence on conventional aeroplanes:

    (a) Varies with aircraft attitude
    (b) Is a predetermined rigging angle
    (c) Is altered using the tailplane

44. Interference drag may be reduced by:

    (a) Fairings at junctions between the fuselage and wings
    (b) Highly polished surface finish
    (c) High-aspect-ratio wings

45. If lift increases, vortex drag:

    (a) Decreases
    (b) Remains the same
    (c) Increases

46. The aspect ratio of a wing may be defined as:

    (a) Chord/span
    (b) Span squared/area
    (c) Span squared/chord

47. In a climb at steady speed, the thrust is:

    (a) Equal to the drag
    (b) Greater than the drag
    (c) Less than the drag

48. Movement of an aircraft about its normal axis is called:

    (a) Rolling
    (b) Pitching
    (c) Yawing

49. The dimension between wing tip and wing tip is known as:

    (a) Wing span
    (b) Wing chord
    (c) Aspect ratio

50. The device used to produce steady flight conditions and relieve the pilot's sustained control inputs is called a:

    (a) Balance tab
    (b) Trim tab
    (c) Servo tab

51. Manometric inputs to the autopilot are typically derived from the:

    (a) Rate gyro
    (b) Centralized air data computer
    (c) Inertial reference system

52. The glide slope (GS) would normally be captured from which pitch mode?

    (a) Heading hold
    (b) Vertical speed from above the GS
    (c) Altitude hold from below the GS

53. The DC polarities of a Dutch roll filter's outputs are greatest when the rate of turn is:

    (a) Maximum
    (b) Minimum
    (c) Zero

54. Autopilots incorporating a pitch-trim control allow the pilot to manually control the aircraft's desired:

    (a) Attitude
    (b) Altitude
    (c) Heading

55. Capturing a VOR radial would normally being achieved from:

    (a) Altitude-hold mode
    (b) Heading-select mode
    (c) Localizer mode

56. Dutch roll occurs with a combination of:

    (a) Yaw and roll disturbance
    (b) Yaw and pitch disturbance
    (c) Roll and pitch disturbance

57. Glide slope (GS) deviation signal gain is reduced to account for:

    (a) Narrowing of the localizer beam
    (b) Decreasing airspeed
    (c) Narrowing of the GS beam

58. Yaw dampers are designed to:

    (a) Allow the Dutch roll frequency to reduce rudder demand
    (b) Allow the Dutch roll frequency to demand rudder in opposition to the disturbance
    (c) Prevent the Dutch roll frequency from demanding rudder in opposition to the disturbance

59. Heading-select mode is incompatible with which other roll mode?

    (a) Localizer on-course
    (b) Localizer capture
    (c) Glide slope capture

60. Glide slope capture is compatible with which other modes?

    (a) VOR capture
    (b) Heading select
    (c) Altitude hold or vertical speed mode

61. Translating tendency in a rotorcraft can be compensated by:

    (a) Increasing thrust from the tail rotor
    (b) Tilting the main rotor to counteract the side-slip
    (c) Increasing torque from the main rotor

62. The collective lever in a rotorcraft is used primarily for:

    (a) Lateral control
    (b) Yaw control
    (c) Thrust and lift

63. Main rotor torque in a rotorcraft is countered by:

    (a) Autorotation
    (b) Coriolis Effect
    (c) Tail rotor thrust

64. During hover, the rotorcraft has a tendency to drift laterally due to the thrust of the:

    (a) Main rotor
    (b) Engines
    (c) Tail rotor

65. Coriolis Effect in a rotorcraft occurs when the centre of mass of the blade moves:

    (a) Away from the axis of rotation, and the rotor velocity increases
    (b) Closer to the axis of rotation, and the rotor velocity increases
    (c) Closer to the axis of rotation, and the rotor velocity decreases

66. When the rotorcraft flies through translational lift, the relative air flowing through the main rotor disc and over the tail rotor disc becomes:

    (a) Less turbulent and hence aerodynamically efficient
    (b) More turbulent and hence aerodynamically efficient
    (c) Less turbulent and hence aerodynamically inefficient

67. Dissymmetry of lift in a rotorcraft is caused by relative airflow over the blades being:

    (a) Added to the rotational relative airflow on the advancing and retreating blades
    (b) Less turbulent and hence aerodynamically efficient
    (c) Added to the rotational relative airflow on the advancing blade, and subtracted on the retreating blade

68. The tail rotor's functions in a rotorcraft are for opposing the:

    (a) Torque created by the main rotor and vertical control of the rotorcraft
    (b) Torque created by the main rotor and directional control of the rotorcraft
    (c) Lift created by the main rotor and directional control of the rotorcraft

69. The cyclic control in a rotorcraft changes the:

    (a) Main rotor's thrust direction
    (b) Tail rotor's thrust direction
    (c) Lift and/or thrust control of the rotorcraft

70. As the rotorcraft accelerates in forward flight, induced flow:

    (a) Increases to near zero at the forward disc area and increases at the aft disc area
    (b) Decreases to near zero at the forward disc area and increases at the aft disc area
    (c) Decreases to near zero at the aft disc area and increases at the forward disc area

71. Attitude rate signals can be derived from:

    (a) Integrating the output from a vertical gyro
    (b) Differentiating the attitude output from a rate gyro
    (c) Differentiating the attitude output from a vertical gyro

72. Linear actuators in a rotorcraft are characterized by being in:

    (a) Parallel with the controls, large displacement, low gain, maximum authority
    (b) Series with the controls, small displacement, high gain, limited authority
    (c) Series with the controls, large displacement, high gain, maximum authority

73. AFCS runaway problems are most likely caused by:

    (a) Vertical gyro offset
    (b) Loss of position feedback from an actuator
    (c) Rate gyro errors

74. AFCS oscillatory problems are most likely caused by:

    (a) Rate gyro errors
    (b) Loss of vertical gyro input
    (c) Loss of position feedback from an actuator

75. Loss of the altitude hold mode is most likely caused by:

    (a) Loss of vertical gyro input
    (b) Rate gyro errors
    (c) Loss of air data inputs

76. An AFCS 'fly-through' feature in a rotorcraft enables the pilot to:

    (a) Make a change to the rotorcraft's flight path by disengaging the AFCS
    (b) Make a change to the rotorcraft's flight path without having to disengage and then re-engage the AFCS
    (c) Fly over a lateral waypoint at a pre-set altitude

77. Gain in a control system is the ratio of:

    (a) Input to output of a control system, which normally translates to the rate of movement of an actuator
    (b) Output to input of a control system, which normally translates to the displacement of an actuator
    (c) Output to input of a control system, which normally translates to the rate of movement of an actuator

78. Cross-coupled feedback in a rotorcraft AFCS is a technique used for:

    (a) Making a change to the rotorcraft's flight path without having to disengage and then re-engage the AFCS
    (b) Tilting the main rotor to counteract the side-slip
    (c) Faster fault detection as well as failure protection

79. The inner loop of the autopilot system is used to make:

    (a) Small flying-control adjustments to counter internal/external disturbances
    (b) Large flying-control adjustments to follow the pilot's control inputs
    (c) Small flying-control adjustments to follow commanded guidance requirements

80. Automatic Stabilization Equipment (ASE) in a rotorcraft uses an attitude reference input to provide:

    (a) Short-term attitude hold function
    (b) Heading hold function
    (c) Long-term attitude hold function

81. Go-around (GA) mode is armed when the aircraft is:

    (a) On the ground roll-out
    (b) Established on the glide slope, and flaps are set to a landing position
    (c) Intercepting the glide slope, and flaps are set to a landing position

82. RVR during an approach informs the pilot of:

    (a) Horizontal visibility on the approach
    (b) Vertical cloud base
    (c) Distance to go along the runway

83. GPS RAIM is achieved by comparing the range estimates made from:

    (a) European geostationary navigation overlay service
    (b) Five satellites
    (c) Six satellites

84. Approach with Vertical Guidance (APV) gives:

    (a) Lateral guidance, but no glide slope
    (b) Lateral and vertical guidance, but to PA criteria
    (c) Lateral and vertical guidance, but not to PA criteria

85. A fail-passive system is able to detect a problem and automatically disconnect the failed channel:

    (a) Leaving the aircraft in an unstable condition
    (b) Without any disturbance to the flight path
    (c) Whilst maintaining automatic control of the aircraft

86. A Category 2 approach is limited to a DH and RVR of:

    (a) 100 feet, 400 metres
    (b) 200 feet, 800 metres
    (c) 50 feet, 200 metres

87. Activating the go-around switch during an ILS approach will:

    (a) Disengage ILS modes, advance the throttles, level the aircraft in pitch and roll
    (b) Disengage ILS modes, advance the throttles, pitch the aircraft nose-up and level the wings
    (c) Maintain guidance using ILS modes, advance the throttles, pitch the aircraft nose-up and level the wings

88. The autothrottle is normally used in conjunction with the autopilot's:

    (a) Roll channel to level the wings during go-around
    (b) Rudder and nose-wheel steering to maintain the aircraft on the runway centre line
    (c) Pitch channel to give a combination of speed and thrust control

89. A paravisual display gives the pilot:

    (a) Fast/slow indications on the ADI
    (b) Visual guidance during the landing rollout
    (c) Vertically directed primary radar information

90. Automatic ILS approaches are usually made by first capturing the:

    (a) Glide slope and then capturing the localizer
    (b) Distance-measuring equipment
    (c) Localizer and then capturing the glide slope

91. Bore-sighting is required on HUD components to ensure that:

    (a) The cockpit environment is NVG compatible
    (b) The desired chromaticity and IR filtering is achieved
    (c) Displayed images are referenced to the earth's horizon and the aircraft's projected flight path

92. Synthetic vision is based on:

    (a) Computer-generated displays that use three-dimensional images
    (b) On-board infrared sensors
    (c) Collimation technology that produces parallel light rays

93. Class B NVIS is when the spectral content starts at:

    (a) 665 nm, allowing wider colour use in avionics systems, with increased sensitivity
    (b) 665 nm, allowing wider colour use in avionics systems, but with reduced sensitivity
    (c) 670 nm, with a secondary peak at 540 nm

94. The HUD flight-path vector or velocity vector symbol indicates:

    (a) If the aircraft is accelerating or decelerating
    (b) The angle of attack
    (c) A predication of the aircraft's flight path

95. Near infrared light wavelengths is made of:

    (a) Visible wavelengths beyond 750 nanometers
    (b) Invisible wavelengths beyond 750 nanometers
    (c) Invisible wavelengths less than 750 nanometers

96. Blooming refers to distorted NVG images surrounded with obscuring halos of light caused by:

    (a) Optically saturation of the goggles
    (b) The primary flying display
    (c) Collimation technology that produces parallel light rays

97. The HUD energy cue display indicates if the aircraft is:

    (a) On the desired track
    (b) Maintaining the desired glide path angle and touchdown point of the runway
    (c) Accelerating or decelerating

98. The spectral content of Class A NVIS starts at:

    (a) 625nm, enabling maximum use of ambient light
    (b) 665nm, enabling maximum use of ambient light
    (c) 625nm, enabling minimum use of ambient light

99. The spectral content above 700–750nm present in starlight is:

    (a) Visible to the unaided human eye
    (b) Not visible to the unaided human eye
    (c) Not visible with NVG

100. When flying in areas or at altitudes where rising terrain may pose a hazard, enhanced vision uses:

    (a) Terrain-alerting database to colourize the landscape
    (b) The primary flying display
    (c) On-board sensors, e.g. FLIR

101. The software used in FBW computers is validated to:

    (a) Level B, in accordance with RTCA D0178/ED-12
    (b) Level A, in accordance with RTCA D0178/ED-12
    (c) Level A, in accordance with RTCA D0160

102. Envelope protection in an automatic system:

    (a) Dynamically controls the flight of the aircraft in order to prevent stalls or overspeed situations
    (b) Dynamically provides the pilot with full-time audible and visual alerting in the event that aircraft is flown outside of its safe limits
    (c) Is an inadvertent, sustained oscillation of the rotorcraft arising from an abnormal handling condition

103. Envelope alerting in an automatic system:

    (a) Dynamically controls the flight of the aircraft in order to prevent stalls or overspeed situations
    (b) Prevents sustained oscillation of the rotorcraft arising from a normal handling condition
    (c) Dynamically provides the pilot with full-time audible and visual alerting in the event that an aircraft is flown outside of its safe limits

104. FBW system architecture is typically based on:

    (a) Multiple computers that have identical software
    (b) Multiple computers that have dissimilar software
    (c) Single-channel computers

105. Rotorcraft pilot-induced oscillation (PIO) is:

    (a) An inadvertent, sustained oscillation of the rotorcraft arising from an abnormal handling condition
    (b) A sustained oscillation of the rotorcraft arising from a normal handling condition
    (c) A full-time audible and visual alerting in the event that an aircraft is flown outside of its safe limits

106. The rotorcraft side-arm controller operates in the conventional sense for:

    (a) Moved vertically for pitch and roll
    (b) Twisted for cyclic pitch and roll
    (c) Cyclic pitch and roll; and is twisted for yaw control

107. Emergency FBW reversion using mechanical linkages is only for:

    (a) Continued safe flight and landing
    (b) Full authority control and protection
    (c) Pilot-induced oscillation

108. Failure of Level A software in certification terms would have:

    (a) Catastrophic consequences
    (b) No safety effect
    (c) Hazardous consequences

109. The probability of complete loss of the FBW system must be

    (a) Remote
    (b) Extremely improbable
    (c) Extremely remote

110. In a FADEC aircraft, the thrust demand from the throttle lever is processed in accordance with control laws to manage the:

    (a) Shaft speed
    (b) Turbine temperature
    (c) Flow of fuel into the engine

111. For a direct-reading compass, the angle of dip is greatest at:

    (a) The equator
    (b) East/west headings
    (c) The magnetic poles

112. The basic 'six pack' configuration is arranged with:

    (a) Top row: airspeed, artificial horizon, altimeter; bottom row: radio compass, direction indicator, vertical speed
    (b) Top row: vertical speed, artificial horizon, altimeter; bottom row: airspeed, radio compass, direction indicator
    (c) Top row: direction indicator, vertical speed, artificial horizon; bottom row: airspeed radio compass, altimeter

113. Acceleration errors in direct-reading compasses are:

    (a) Minimum on east/west headings
    (b) Maximum on north/south headings
    (c) Maximum on east/west headings

114. In reversionary mode, all-important PFD flight information is shown on the:

    (a) Electronic flight bag
    (b) Remaining display
    (c) AHRS

115. The gyro compass synchronizing indicator shows when the gyroscopic and magnetic references are:

    (a) Not in agreement
    (b) In agreement
    (c) Powered off

116. The PFD is a single vertical instrument that replaces the existing

    (a) Altimeter/VSI/ASI
    (b) Electronic Flight Bag
    (c) Attitude Indicator and Heading Indicator/HSI

117. Magnetic interference caused by the aircraft's structure and components is called:

    (a) Deviation
    (b) Variation
    (c) Magnetic dip

118. Class 1 EFB systems are typically:

    (a) Standard commercial off-the-shelf devices
    (b) Fully certified and integrated with aircraft systems
    (c) Viewable to the pilot during all phases of flight

119. The flux gate sensor is a pendulous device, free to move in response to:

    (a) Pitch and yaw, but fixed in roll
    (b) Pitch and roll, but fixed in yaw
    (c) All three axes – pitch, roll and yaw

120. The 'flying T' configuration is arranged with the:

    (a) Artificial horizon/attitude indicator in the top/centre, air-speed indicator (ASI) to the right, altimeter to the left and directional gyro underneath
    (b) Artificial horizon/attitude indicator in the top/centre, air-speed indicator (ASI) to the left, altimeter to the right and directional gyro underneath
    (c) Directional gyro in the top/centre, air-speed indicator (ASI) to the left, altimeter to the right and artificial horizon/attitude indicator underneath

Appendix C: Answers to questions

CHAPTER 1

1. a
2. c
3. b
4. a
5. c
6. b
7. c
8. a
9. a
10. b

CHAPTER 2

1. b
2. a
3. a
4. c
5. b
6. b
7. a
8. c
9. a
10. b

CHAPTER 3

1. b
2. a
3. c
4. a
5. b
6. c
7. a
8. a
9. b
10. c

CHAPTER 4

1. c
2. a
3. c
4. b
5. a
6. c
7. a
8. a
9. b
10. b

CHAPTER 5

1. b
2. a
3. a
4. b
5. c
6. a
7. b
8. b

9. b
10. b

CHAPTER 6

1. a
2. c
3. b
4. a
5. c
6. b
7. a
8. c
9. a
10. b

CHAPTER 7

1. a
2. c
3. b
4. a
5. c
6. b
7. a
8. c
9. a
10. b

CHAPTER 8

1. b
2. c
3. a
4. a
5. b
6. a
7. c
8. b
9. a
10. c

CHAPTER 9

1. b
2. a

3. c
4. c
5. b
6. a
7. c
8. b
9. a
10. b

CHAPTER 10

1. b
2. a
3. b
4. c
5. a
6. b
7. a
8. c
9. a
10. c

CHAPTER 11

1. b
2. a
3. b
4. c
5. b
6. a
7. b
8. c
9. b
10. c

CHAPTER 12

1. c
2. a
3. b
4. c
5. b
6. a
7. c
8. a
9. b
10. c

CHAPTER 13

1. b
2. a
3. c
4. b
5. a
6. c
7. a
8. a
9. b
10. c

REVISION QUESTIONS

1. b
2. c
3. a
4. c
5. b
6. a
7. c
8. a
9. c
10. a
11. a
12. c
13. b
14. a
15. c
16. b
17. c
18. a
19. a
20. b
21. b
22. a
23. c
24. a
25. b
26. c
27. a
28. a
29. b
30. c
31. a
32. c
33. b
34. a

35. c
36. b
37. a
38. c
39. a
40. b
41. a
42. c
43. b
44. a
45. c
46. b
47. a
48. c
49. a
50. b
51. b
52. c
53. a
54. a
55. b
56. a
57. c
58. b
59. a
60. c
61. b
62. a
63. c
64. c
65. b
66. a
67. c
68. b
69. a
70. b
71. b
72. a
73. b
74. c
75. a
76. b
77. a
78. c
79. a
80. c
81. b
82. a
83. b
84. c

85.	b		103.	c
86.	a		104.	b
87.	b		105.	a
88.	c		106.	c
89.	b		107.	a
90.	c		108.	a
91.	c		109.	b
92.	a		110.	c
93.	b		111.	c
94.	c		112.	a
95.	b		113.	c
96.	a		114.	b
97.	c		115.	a
98.	a		116.	c
99.	b		117.	a
100.	c		118.	a
101.	b		119.	b
102.	a		120.	b

Index

Printed in the United States
by Baker & Taylor Publisher Services